The HERBARIST

The Early Years

1935–39

A Publication of
The Herb Society of America

Printed by CreateSpace.com

THE HERB SOCIETY OF AMERICA
9019 Kirtland-Chardon Road
Kirtland, OH 44094
www.herbsociety.org

Originally published 1935–39 by
The Herb Society of America.

ISBN: 978-0-9979242-0-6

Dedicated to our founders:

Florence Brown Bratenahl
Harriet Addams Brown
Anne Shirk Burrage
Adeline Philbrick Cole
Ella Hulst Greenslet
Corinna Searle Mitchell
Frances Torrey Norton

FOREWORD

On August 23, 1933 seven women, who summered on Boston's North Shore, met to organize a society dedicated to the study of herbs. The ladies were well-educated and ambitious. All were avid gardeners. With an eye to the future, they called themselves The Herb Society of America, and adopted the motto: For Use and for Delight.

The next year they made plans to enter the Massachusetts Horticultural Society's 1935 Boston Spring Flower Show with a display that would recreate a Seventeenth Century Still Room. The enormous amount of research required took them most of 1934, but by the following spring they were ready. It was a notable first effort, and the judges awarded them first place.

Following their success, founder Adeline Cole wrote an article about their exhibition for a bulletin to be produced for The Society's members. It debuted in 1935 as *The Herbarist*, and has been The Society's annual publication ever since. Only 500 were printed that first year, and each sold for seventy-five cents. The second year, the price rose to one dollar, and a thousand were published. To defray the cost of printing, herb-related shops and gardens purchased ads, many of which contained charming hand drawings.

The early years of *The Herbarist* offer articles written by the founders and first members, and continue to be in great demand. As it has become increasingly difficult to obtain the original publications, it was decided to reproduce the years 1935–1939 in their entirety. We hope you will enjoy them.

The HERBARIST

A Publication of the

Herb Society of

America

For Use and for Delight

BOSTON

1935

Published by the Herb Society of America
Horticultural Hall
300 Massachusetts Avenue
Boston, Massachusetts

F O R E W O R D

WHY AN HERB SOCIETY?

WHY should another society be admitted into the circle of special groups? Is it the fad of the moment for garden club members who find a new interest in making large collections, creating herb gardens in interesting designs and sapping the juice from the old herbals?

What is there in herbs to arrest the attention of a small group of people scattered over the United States, to make them feel the need for a clearing house for their interest?

As horticulture has developed our flowers into newer and better varieties of color and form, the herb, having only fragrance to offer, has been pushed from the flower garden, leaving but a trail of Thyme between the rock crevices. The kitchen garden, too, has lost its varied flavors, being supplanted on our pantry shelves by prepared poultry dressing and a bottle of Tarragon vinegar.

The increasing interest in herbs is changing the status of these orphans of the garden; newer and better varieties, both for fragrance and flavor, are being cultivated, and we see the herb established in gardens of its own or introduced as a minor note in the flower garden.

During this process of development there is a very real reason why the growers who are interested in the cultivation and increased use of herbs should have such a clearing house as this society to which to bring their problems, experiments and achievements.

Still houſe

Reproduced from
A New Orchard and Garden
by W. Lawson, 1617

NATURE'S BAGS OF SCENT

EDGAR ANDERSON

A SIMPLE hand lens, a folding pocket hand lens of glass and nickel, is a window into another world in which the obscure details of our naked-eye existence become magnified until they reveal their significance, variety and charm. Through a hand lens the felty surface of a mullen leaf becomes a silver forest of fantastic interlocking branches; the center of a violet is seen as an ornamented throne draped with richly-patterned velvet textiles of purple, white and gold. To the herbarist this charming world is more than a delight to the senses; it is of some practical importance in the production of scent and savor. Some of the aromatic substances to which our herbs owe their usefulness are carried deep inside the plant, but many are borne either on the surface or are even held above it in curious little globes or saucers. These scent-producing organs are known to botanists as "glands" or "glandular hairs." For the most part they take the form of smooth, secretory surfaces or of microscopic sacs.

In the rosemary the scent glands are most easily found on the green calyx just below the flower. They are thickly sprinkled over the surface, and through a hand lens shine like opalescent pearls against a background of green velvet. In wormseed the curious little glands which carry the essential principle of the herb are found deep within the flower itself. Each gland looks like a little foot attached at the ankle. The whole gland is quite transparent and within it can be seen a floating drop of a clear, golden oil. In hops the bitter principle is produced in fantastic, trumpet-shaped glands borne near the flower, particularly on the contorted, leaf-like structures which botanists have named "bractioles." When the hop flowers first open the trumpets are empty, though they glisten brightly within. As the flowers mature the golden trumpets fill with the golden liquid, which darkens and hardens as it dries. When the hops are finally ready to be harvested, the trumpets have become cornucopias piled with a golden-brown exudate. They snap off from the plant at a brush of the hand; each one, to the naked eye, is a golden-brown grain of dust. It is hundreds of millions of these little trumpets which form the golden-brown "hop-meal" known to brewers and hop merchants.

If you are becoming an herbarist do buy a simple hand lens. One which magnifies eight to ten times is quite strong enough. Through it you will find a whole new, interesting world awaiting you. Through it you can amass for yourself another sort of herb lore. You will bring to light hidden beauties in the most humble garden herbs and understand the production of scent and savor with the precision of a master herbarist.

ARNOLD ARBORETUM

February 1935

THE FOLLOWING PLATES from the "Histoire Naturelles des Lavandes" may well be preceded by the same words with which Frédéric Charles Jean, Baron Gingins de Lassaraz concluded his original introduction in 1826: "La Lavande réunit ainsi l'utilité à l'agrément et si les gens du monde étaient plus curieux de connaître les choses dont ils usent chaque jour, ils comprendraient qu'on ait pû écrire l'Histoire des Lavandes."

Tab. I.

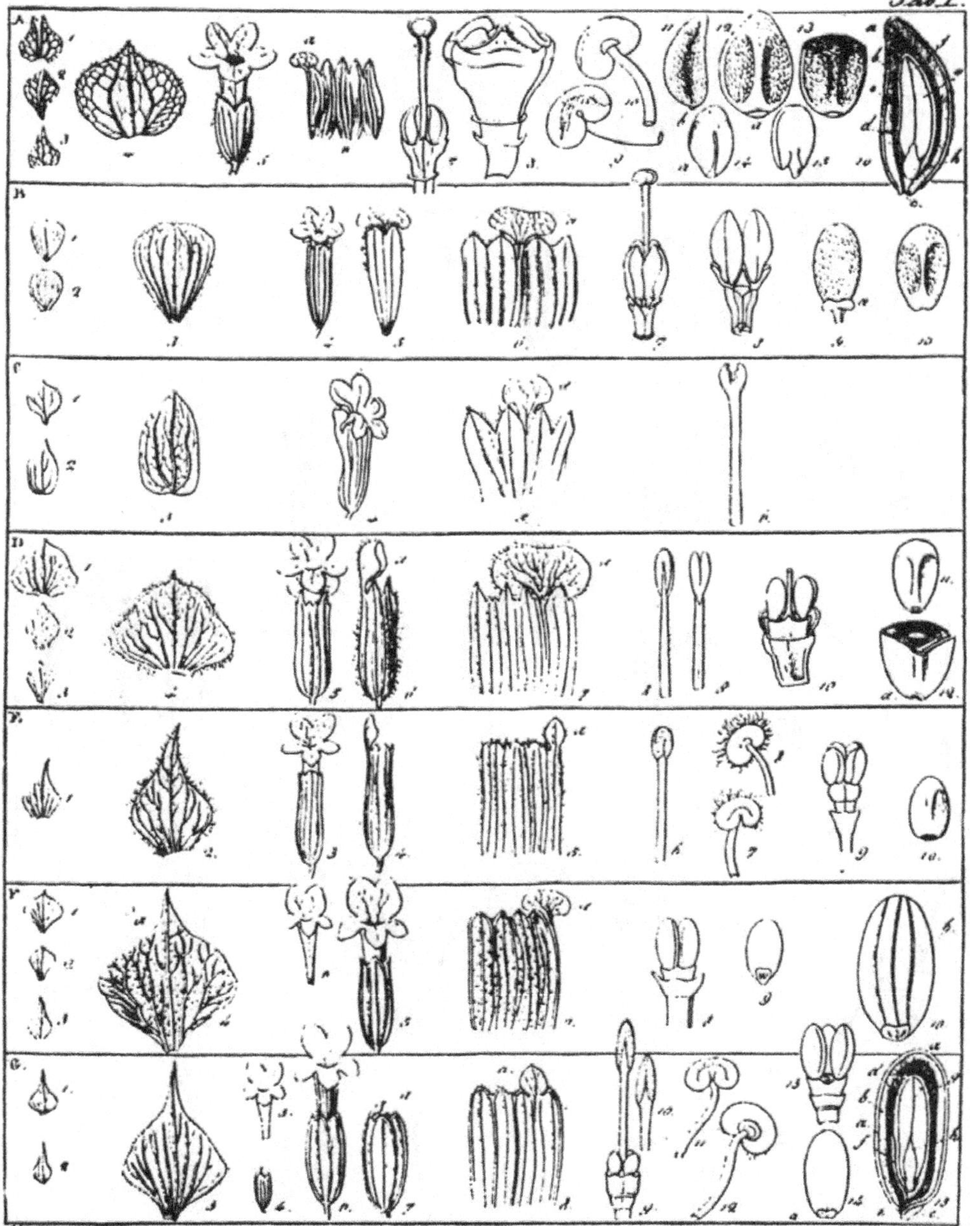

Analyses des espèces

Tab. II.

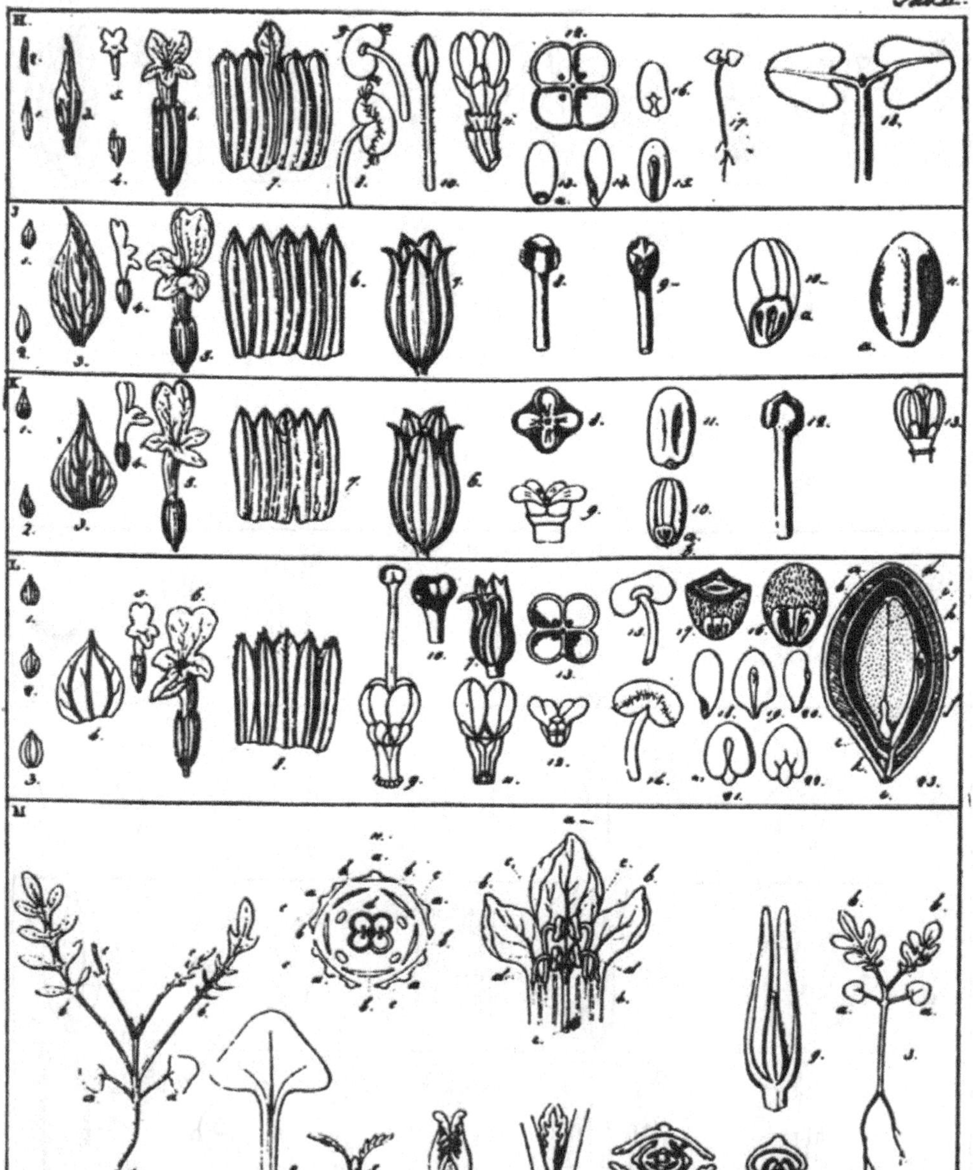

Analyses des espèces

Tab. III.

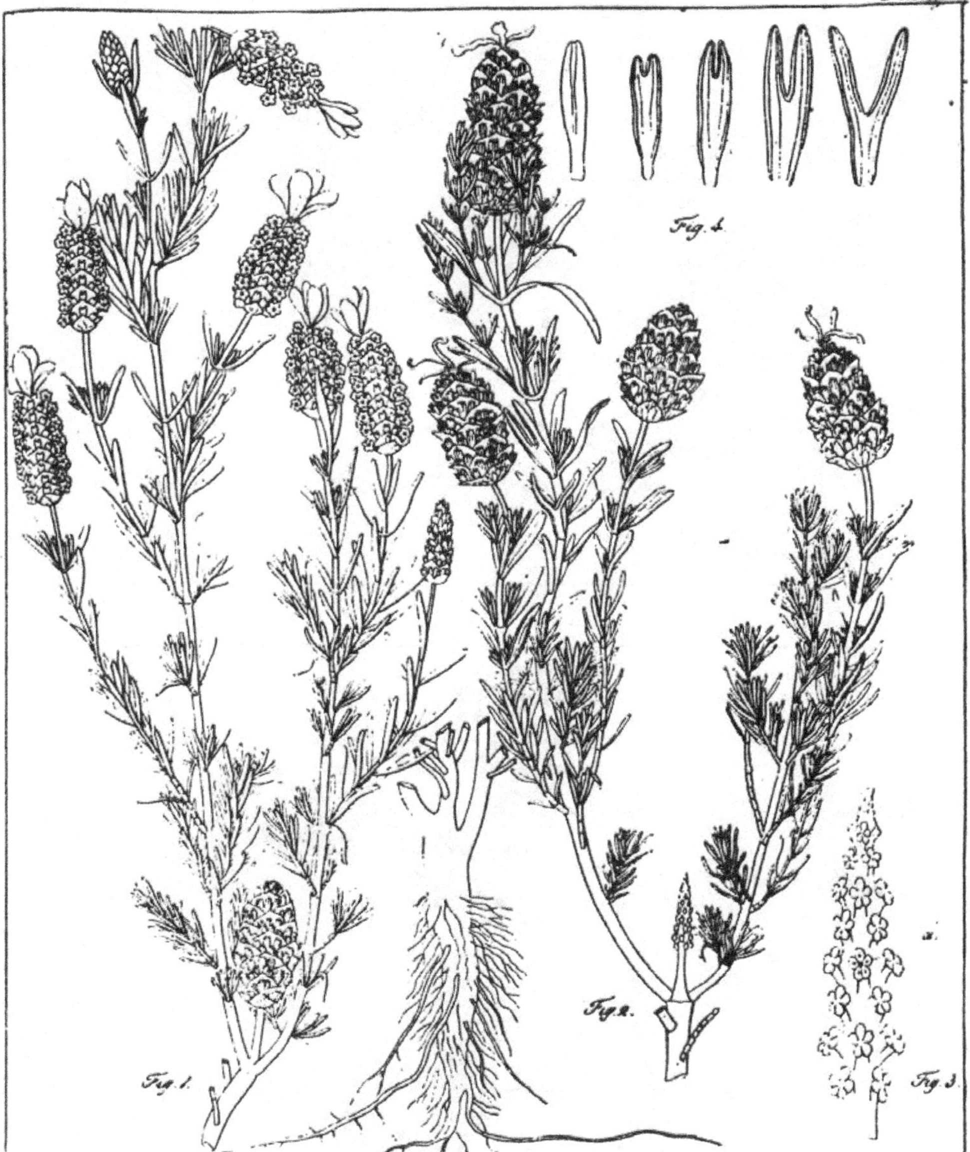

Lavandula Stæchas, Linn.

Lavandula pedunculata, Cavan (Fig. 1)
Lavandula viridis, L'Her (Fig. 2)

Lavandula dentata, Linn.
Fig. 1. var. vulgaris. Fig. 2. var. B. Balearica

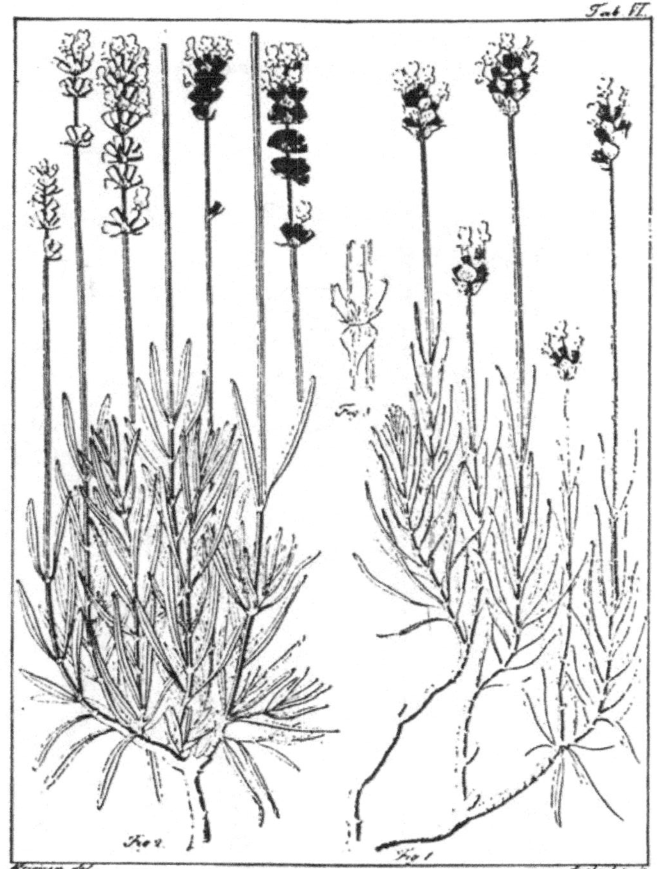

Lavandula Pyrenaica D. C. (*Fig. 1*)
Lavandula Vera D. C. (*Fig. 2, 3*)

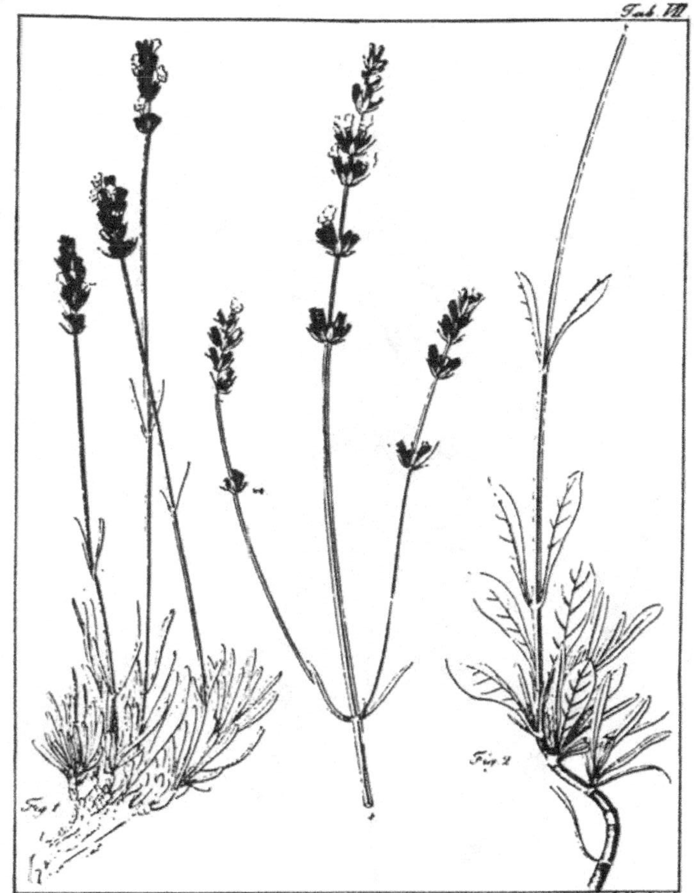

Lavandula Spica D. C.
Fig. 1, a vulgaris. Fig. 2, B. ramosa

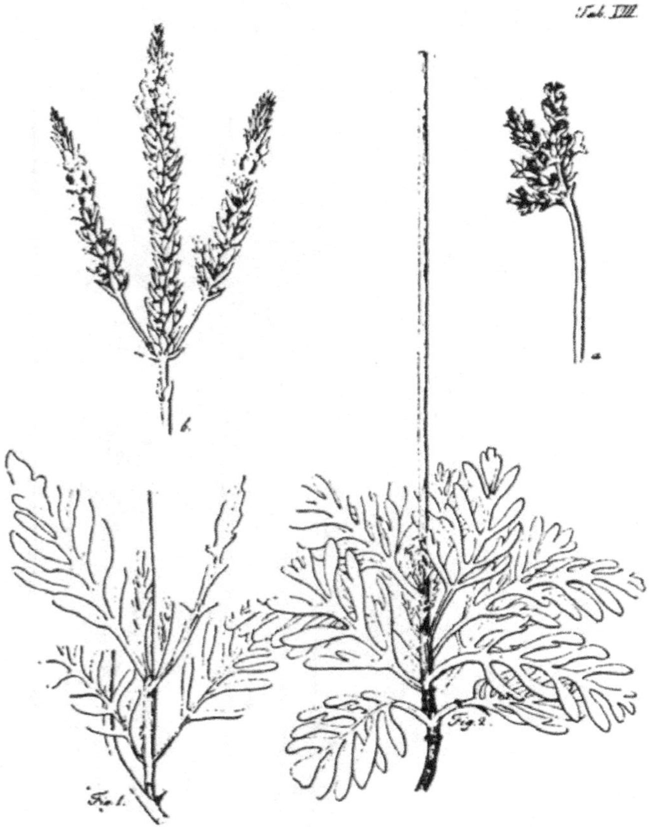

Lavandula pinnata, Linn. fil.
Fig. 1, var. a. Fig. 2, var. b.

Lavandula corono pifolia, Poir.

Tab. X

Lavandula abrotanoïdes, Lam.

Tab. XI

Lavandula multifida, Linn.

THE 17th CENTURY STILL-ROOM

ADELINE P. COLE

The knowledge of
Stilling is one pretty feat
The waters be wholesome
The charges not greate.
—*Tusser's "500 Points of Good Husbandry,"* 1573.

THE good house-wife of the 17th century was expected to be very wise in the preparation and use of the herbs she grew; in large country houses a room was set apart for the lady of the manor to carry on these house-wifely arts: this room was called the still-room, and its story is almost hidden in the closely-woven warp of the domestic life of the period. To the 20th century such a room is a sort of mystery, as it has no counterpart in our life; the still-room was a sort of composite, partly a laboratory, a medicine closet, and somewhat a storeroom and pantry.

Its location was not fixed; we may pore over old plans, read into the domestic life of the people, yet fail to find that it had a definite place in the manor house, like the kitchen or dairy; it was there, a part of every country house, sometimes even on the second floor,[1] next the bedrooms; sometimes, as in the drawing of William Lawson,[2] in a tiny detached house. But in whatever place we find it, it was here that the lady of the manor directed her maids in the composition of the laborious recipes by which she produced the domestic remedies for her household, and the poor in her parish; here she dried the herbs for the flavors for the kitchen, and distilled the sweet waters, scents and toiletteries in which she delighted.

We get a very good picture of the activities in the still-room from the old still-room books, in which the cherished recipes were kept. Some of these books are preserved in MS. in the libraries in England, others are still in possession of descendants of the writers; fortu-

[1] Hawstead, Surrey.
[2] *A New Orchard and Garden* by W. Lawson, 1617.

nately, a few have been made more available by publication.[3]

The still-room book has a character all its own; it is more than a cook book, less than an herbal, but it contains the accumulated knowledge of the family of the virtues and preparation of the plants, which formed such an important part of the domestic economy of the day. How these 17th century women cherished these recipes from their friends! What joy to plain Elizabeth Wainwright to get the recipe for "Lady Allen's Water"; it reads like an herbal, with its list of thirty-three herbs to be steeped in white wine and brandy, and then distilled.

In addition to these MS. books, several contemporary works were printed "contayning the vertious knowledge . . . which ought to be in any compleat housewife, of what degree or calling soever."[4]

These also were collections of recipes from relatives and friends, mixed with the writer's own philosophy as to a woman's duties.

An early "Approved Book" was the *English House-Wife* of Gervase Markham; to begin with, Markham would have this ideal woman skilled in the preparation of medicines for the health of her household; then she must have knowledge of all sorts of herbs belonging to the kitchen, which skill she must get by her own labor and experience: "When our English house-wife is exact in these rules she shall then sort her mind to other secrets. Therefore first I would have her furnish herself with a good still for the distillation of sweet waters. . . . Then she shall know that the best waters for the smoothing the skin, and keeping the face delicate and amiable are those distilled from strawberries, flowers of lillies, etc." With this summary of the duties of an English house-wife, and his large collection of recipes, Gervase Markham gives a good index of the activities in the still-room of the 17th century.

---

[3] *A Proper Newe Booke of Cokery.* From a Ms. left to Corpus Christi College, Cambridge, by Archbishop of Canterbury.
Customs of Yardley Hastings.
Arcana Fairfaxiana. Facsimile of Mary Cholmeley Still-Room Book.
Plain Plantain. Madam Susanna Avery Ms. Still-Room Book.

[4] *Jewel House of Art and Nature* by Sir Hugh Platt.
Tusser's *500 Points of Good Husbandry.*
Country House-Wife's Garden by William Lawson, 1617.
English House-Wife by Gervase Markham, 1637.

A LIST FOR AN OLD ENGLISH

WORT GARDEN

ELLEN GREENSLET

WORT is a word that has been in the English language since Anglo-Saxon times, meaning originally a root. Later it was transferred to any kind of plant, whether growing wild, or cultivated in a garden. Although a wort is not necessarily an herb, as we use the term today, yet long before Chaucer's time, in those early days when "herb" carried a less specialized significance than it does at present, the two words were practically synonymous, and so remained until the early 17th century. Indeed, throughout their entire history, most of the worts have come well within our modern usage and definition of "herb."

During the middle decades of the 17th century "plant," as a general term, took the place of the earlier "wort" and "herb," and "wort" gradually fell into disuse, except in the names of certain old plants, where it had been used as a suffix, or as a "second element," as some authorities prefer to speak of it. Many of these names are now used in everyday speech only in out-of-the-way corners of Great Britain, though a few have lingered on as the common names of certain genera and species. Except for this latter usage, many of these vernacular wort-names are found today only in old herbals and gardening books, in folk literature, and in scholarly treatises and lists, where their origins and evolution may be studied.

Hunting for these early worts is an interesting form of indoor botanizing; but it is a study which, to be entirely accurate in its results, would call for highly specialized and detailed research, which would necessarily develop into a lengthy article. The following list is the result of a weeding-out process, fifty or more worts having been discarded as ineligible either because of their modern "book names," or because they were not known in Jacobean England. There is, however, sufficient material to serve as a nucleus for the study—either indoors or out—of these old plants of the country-side, all of them valued or loved because of their association in Chaucer's and Shakespeare's time.

"A GALLANT HERB"—Waywort or Pimpernel

(*Anagallis Arvensis, Linn.*)

" There are twoe kindes, differing in the flower, for that which hath an azure flower is called the femall, but that of a Phœnician colour is called the Male."

DIOSCORIDES, *Englished by John Goodyer,* A.D. 1655.
Edited and First Printed, A.D. 1933.
By Robert T. Gunther. Oxford, 1934.

This drawing of one of the oldest known worts, as reproduced in Mr. Gunther's edition, very much reduced in size from the original, is found in the Aniciæ Julianæ, a beautifully illustrated Byzantine manuscript of about 512 A.D., now preserved in the Vienna Library. This is one of several of its illustrations based upon originals by Crateuas, a celebrated Greek rhizotomist of the second century B.C., who described, and drew from nature, a number of medicinal plants. A few of those descriptions are known to us, having been quoted by Dioscorides in his Greek Herbal written in the first century A.D.

The bare bones of a list show so little result for all the pleasant and far-reaching ramifications of reading that have gone into its make-up. But it has been a peculiarly rewarding study in many ways, and it is hard to imagine a pleasanter bypath to follow when winter gardening has to take the place of one's outdoor summer activities.

Worts have a confusing habit of sharing their names with each other, one plant, such as the Greater Celandine, often having several wort synonyms. Any one of these, in its turn, may apply to two or three entirely distinct and quite unrelated plants. Bloodwort is an example of this, being a name used for White Clover, Centaury, Yarrow, Smartweed and Tormentil. There is also great divergence of opinion as to the Latin names of certain species. In the main, unless the balance of opinion seems to be against it, Mr. Britten's nomenclature has been followed.

Bibliography

Many interesting books, both old and new, have been consulted, but the list cannot be given for lack of space. The following six have been used constantly for reference:

Earle, J.: *English Plant Names from the Tenth to the Fifteenth Century.* Oxford, 1880.

Prior, R. C. A.: *Popular Names of English Plants.* London, 1870.

Britten, J., and Holland, R.: *A Dictionary of English Plant Names.* London, 1886.

Bentham and Hooker: *Handbook of the British Flora,* 1924 Edition.

Mrs. Leyel's and Mrs. Grieve's *Herbal:* A number of worts are considered. New York, 1931.

Horwood, A. R.: *British Wild Flowers,* 6 volumes, 1919. This work has been especially valuable. It treats of a number of plants from the standpoint of ecology and bionomics. Excellent "Life Histories" of a number of worts are found among the species described. London, 1919.

A LIST FOR AN OLD ENGLISH WORT GARDEN

Wort	Common name	Botanical name
Adderwort	Snakeweed	*Polygonum Bistorta, L.*
All Saints' Wort	Tutsan	*Hypericum Androsæmum, L.*
Axwort	Axweed	*Coronilla Varia, L.*
Bairnwort	English daisy	*Bellis perennis, L.*
Banewort	Deadly nightshade	*Atropa belladonna, L.*
Banwort	Sweet violet	*Viola odorata, L.*
	Wallflower	*Cheiranthus cheiri, L.*
	Centaury	*Centaurium umbellatum, Gilib.*
	Heart'sease	*Viola tricolor, L.*
Barrenwort	Bishop's hat	*Epimedium alpinum, L.*
Bearwort	Meu or spignel	*Meum athamanticum, L.*
Beewort	Sweet flag	*Acorus calamus, L.*
Birthwort	Birthwort	*Genus Aristolochia*
		Aristolochia clematitis
Bishopswort	Herb Gerard	*Ægopodium podagraria, L.*
	Wood betony	*Stachys Betonica, Benth.*
	Water mint	*Mentha aquatica, L.*
Bitterwort	Gentian	*Gentiana amarella, L.*
	Dandelion	*Taraxacum officinale, Weber*
Blackwort	Boneset	*Symphytum officinale, L.*
	Comfrey	
Bladderwort	Hooded water milfoil	*Utricularia vulgaris, L.*
Blawort	Harebell	*Campanula rotundifolia, L.*
	Cornflower or Blue-bottle	*Centaurea cyanus, L.*
Bledewort	Red poppy	*Papaver rhoeas, L.*
Blisterwort	Celery-leaved Crow-foot	*Ranunculus sceleratus, L.*
Blodewort	Waterpepper	*Polygonum hydropipa, L.*
Bloodwort	Whiteclover	*Trifolium repens, L.*
	Smartweed	*Polygonum hydropiper, L.*
	Yarrow or Common Milfoil	*Achillea Millefolium, L.*
	Centaury	*Erythræa Centauria, Pers.*
	Tormentil	*Potentilla erecta, Hampe*
Bogwort	Cranberry	*Vaccinium Oxycoccus, L.*
Bridewort	Meadowsweet	*Spiræa Ulmaria, L.*
Brimstonewort	Milk parsley	*Peucedanum palustre, L.*
Broadwort	Water figwort	*Scrophularia aquatica, L.*
Broomwort	Water betony	*Scrophularia aquatica, L.*
	Pennycress	*Various species of Thlaspi*
Brosewort	Common henbane	*Hyoscyamus niger, L.*
Brotherwort	Wild thyme	*Thymus serpyllum, L.*
Brownwort	Self-heal	*Prunella vulgaris, L.*
	Water figwort	*Scrophularia aquatica, L.*
Bruisewort	Yellow-horned poppy	*Glaucium flavum, Crantz.*
	Comfrey	*Symphytum officinale, L.*
	Common daisy	*Bellis perennis, L.*
Bullwort	Water betony	*Scrophularia aquatica, L.*
Burwort	Lesser Celandine	*Ranunculus ficaria, L.*
Butterwort	Common Butterwort	*Pinguicula vulgaris, L.*
Cankerwort	Dandelion	*Taraxacum officinale, Weber*
Chafewort	Cudweed	*Filago germanica, L.*
Chickenwort	Chickweed	*Stellaria media, L.*
Christ's-wort	Christmas rose	*Helleborus niger, L.*
Churchwort	Pennyroyal	*Mentha pulegium, L.*
Clovewort	Buttercup or crow-foot	*Ranunculus acris, L.*
Cnop-wort	Knot-weed	*Centaurea nigra, L.*
Cocowort	Shepherd's purse	*Capsella Bursa-pastoris, Medik.*
	Avens or Herb Bennett	*Geum urbanum, L.*
Colewort	Cabbage or Kale	*Brassica oleracea, L.*
Colickwort	Parsley Piert	*Alchemilla arvensis, L.*
Coughwort	Coltsfoot	*Tussilago Farfara, L.*
Cow's-wort	Marsh red rattle	*Pedicularis palustris, L.*
Cowthwort	Motherwort	*Leonurus cardiaca, L.*
Crosswort	Crosswort	*Galium cruciata, Scop.*
Cudwort	Cudweed	*Filago germanica, L.*
Culverwort	Columbine	*Aquilegia vulgaris, L.*
Curdwort	Yellow bedstraw	*Galium vernum, L.*
Danewort	Dwarf elder	*Sambucus ebulus, L.*
Deadwort	Dwarf elder	*Sambucus ebulus, L.*
Dragon-wort	Snake-root	*Polygonum Bistorta, L.*

Dropwort	Dropwort	*Spiræa filipendula, L.*
Edder-wort	Dragonwort	*Arum Dracunculus, L.*
Elfwort	Elecampane	*Inula Helenium, L.*
Ers-wort	Mouse-ear hawkweed	*Hieracium pilosella, L.*
Fellow-wort	Herb Robert	*Geranium robertianum, L.*
Felonwort	Greater Celandine	*Chelidonium majus, L.*
	Bittersweet	*Solanum dulcamara, L.*
Feltwort	Aaron's rod	*Verbascum thapsus, L.*
Felwort	Autumn gentian	*Gentiana amarella, L.*
Feverwort	Centaury	*Erythræa Centaurium, Pers.*
Field-wort	Autumn gentian	*Gentiana amarella, L.*
Figwort	Water figwort	*Scrophularia aquatica, L.*
		Genus Scrophularia
	Lesser Celandine	*Ranunculus ficaria, L.*
Filewort	Field cudweed	*Filago minima, Pers.*
Filwort	Centaury	*Erythræa Centaurium, Pers.*
Fingerwort	Foxglove	*Digitalis purpurea, L.*
Fleawort	Ploughman's spikenard	*Inula conyza, DC.*
Flewort	Ribwort	*Plantago lanceolata, L.*
Flirtwort	Feverfew	*Pyrethrum parthenium, L.*
Flixwort	Flixweed	*Sisymbrium sophia, L.*
Flowkwort	Marsh Pennywort	*Hydrocotyle vulgaris, L.*
Foalswort	Coltsfoot	*Tussilago Farfara, L.*
Frogwort	Purple orchis	*Orchis mascula, L.*
Gallwort	Common Toadflax	*Linaria vulgaris, L.*
Garlick-wort	Jack-by-the-hedge	*Alliaria officinalis, Andrz.*
Gipsy-wort	Water horehound	*Lycopus europæus, L.*
Glasswort	Saltwort	*Salicornia herbacea, L.*
Goosewort	Silverweed	*Potentilla anserina, L.*
Gout-wort	Herb Gerard	*Ægopodium podagraria, L.*
Grape-wort	Herb Christopher	*Actea spicata, L.*
	Common Bryony	*Bryonia dioica, Jacq.*
Hammer-wort	Pellitory-of-the-wall	*Parietaria officinalis, L.*
Hazelwort	Asarabacca	*Asarum Europæum, L.*
Heart-wort	Common melilot	*Melilotus officinalis, Wild*
Hartwort	Hartwort	*Tordylium maximum, L.*
Hertwort	Hertwort	*Various species of Aristolochia*
	Ash tree	*Fraxinus excelsior, L.*
Lungwort: Bullock's, Cow's, Clown's	Great Mullein	*Verbascum thapsus, L.*
Madderwort	Madwort	*Genus Alyssum*
Maderwort	Wormwood	*Artemisia absinthium, L.*
Madwort	Madwort	*Genus Alyssum*
		Asperugo procumbens, L.
Marshwort	Cranberry Bogwort	*Oxycoccus palustris, Pers., L.*
Masterwort	Masterwort	*Peucedanum ostruthium, Koch.*
	Pellitory-of-Spain	*Imperatoria ostruthium, L.*
	Herb Gerard	*Ægopodium podagraria, L.*
Maudlin wort	Ox-eye daisy	*Chrysanthemum leucanthemum, L.*
May-wort	Golden Crosswort	*Galium cruciata, L.*
Meadwort, Medewort, Meadow-wort	Meadowsweet	*Spiræa Ulmaria, L.*
Mekilwort	Deadly nightshade	*Atropa belladonna, L.*
Michelwort	White Hellebore	*Veratrum album, L.*
Milkwort	Common Milkwort	*Polygala vulgaris, L.*
Mingwort	Mugwort	*Artemisia vulgaris, L.*
Money-wort	Creeping Jenny	*Lysimachia nummularia, L.*
	Bog pimpernel	*Anagallis tenella, L.*
Moonwort	Moon fern	*Botrychium lunaria, L.*
	Greater Stitchwort	*Stellaria holostea, L.*
Moor-wort	Wild rosemary	*Drosera rotundifolia, L.*
Motherwort	Mugwort	*Artemisia vulgaris, L.*
	Motherwort	*Leonurus cardiaca, L.*
Mothwort	Immortelle	*Helichrysum stoechas, DC.*
Mudwort	Mudweed	*Limosella aquatica, L.*
Mugwort	Mugwort	*Artemisia vulgaris, L.*
Nailwort	Rue-leaved saxifrage	*Saxifraga tridactylides, L.*
	Whitlow Grass	*Draba Verna, L.*
Navel-wort	Wall pennywort	*Cotyledon umbilicus, L.*
Nettlewort	Nettlewort	*Whole order of Urticaceæ*
Nipplewort, Nipwort	Nipplewort	*Lapsana communis, L.*
Oderwort	"The Herb Dragance" Snakeweed	*Polygonum Bistorta, L.*
Opopanewort	Marsh Woundwort	*Stachys palustris, L.*

Herringswort	Common Bugle	*Ajuga reptans, L.*
Hillwort	Wild thyme	*Thymus serpyllum, L.*
	Pennyroyal	*Mentha pulegium, L.*
Hip-wort	Common navelwort	*Cotyledon umbilicus, L.*
	Pennywort	
Hoarwort	Cudweed	*Filago germanica, L.*
Horewort		
Homewort	Common houseleek	*Sempervivum tectorum, L.*
Honewort	Honewort	*Trinia glauca, Dumort.*
	Corn parsley	*Petroselinum segetum, Koch.*
Honeywort	A hawkweed	*Hieracium cerinthoides, L.*
	Maywort	*Galium cruciata, Scop.*
Hoodwort	Skull-cap	*Scutellaria galericulata, L.*
Hornwort	Hornweed	*Ceratophyllum demersum, L.*
Iron-wort	Mayweed	*Anthemis cotula, L.*
Ivy-wort	Ivy-leaved toadflax	*Linaria Cymbalaria, Mill*
*Juph-wort	not identified—used in Markham's "English Housewife." 1637.	
Kelpwort	Prickly Saltwort	*Sakola kali, L.*
Kenningwort	Greater Celandine	*Chelidonium majus, L.*
Kernelwort	Figwort	*Scrophularia nodosa, L.*
Kidneywort	Navelwort	*Cotyledon umbilicus, L.*
Knight's-wort	Water-soldier	*Stratiotes aliodes, L.*
	Woundwort	
	Knight's Pondwort	
Knot-wort	Knotgrass	*Polygonum aviculare, L.*
Lagwort	Butterbur	*Tussilago Petasites, L.*
Langwort	White or False	
Lingwort	Hellebore	*Veratrum album, L.*
Latherwort	Soapwort	*Saponaria officinalis, L.*
Leadwort	Common Leadwort	*Plumbago europæa, L.*
Leechwort	Ribwort Plantain	*Plantago lanceolata, L.*
Lichwort	Pellitory-of-the-wall	*Parietaria officinalis, L.*
Limewort	Viscid Campion	*Lychnis viscaria, L.*
Limpwort	Brooklime	*Veronica beccabunga, L.*
Lithewort	Wayfaring tree	*Viburnum lantana, L.*
Liverwort	Common Agrimony	*Agrimonia Eupatoria, L.*
		Marchantia polymorpha, L.
Longwort	Pellitory-of-Spain	*Anacyclus pyrethrum, L.*
Lousewort	Marsh red rattle	*Pedicularis palustris, L.*
Lungwort	Lungwort	*Pulmonaria officinalis, L.*

Paisywort	Cowslip	*Primula veris, L.*
Peachwort	Persicaria	*Polygonum persicaria, L.*
Pearlwort	Pearlwort or Spurry	*Genus Sagina*
Pennywort	Navelwort	*Cotyledon umbilicus, L.*
	Ivy-leaved toadflax	*Linaria Cymbalaria, L.*
Pepperwort	Peppergrass	*Lepidium campestre, Br.*
Pestilence-wort	Butterbur	*Petasites officinalis, Moench.*
Peter's wort	Hard Hay	*Hypericum tetrapterum, L.*
Pewter wort	Rough horse-tail	*Equisetum hyemale, L.*
Pilewort	Lesser Celandine	*Ranunculus ficaria, L.*
Pipe-wort	Pipewort	*Family Eriocaulaceæ*
Pond-wort	Water-soldier	*Stratiotes aloides, L.*
Pricksong-wort	Honesty	*Linaria biennis, L.*
Purplewort	Marsh cinque-foil	*Comarum palsture, L.*
Quidwort	Cudweed	*Filago germanica, L.*
Quillwort	Quillwort	*Isoetes lacustris, L.*
Quincywort	Squinancy-wort	*Asperula cynanchica, L.*
Ragwort	Ragweed	*Senecio Jacobæa, L.*
Ribwort	Ribwort	*Plantago lanceolata, L.*
Rybwort		
Rupture-wort	Rupture-wort	*Herniaria glabra, L.*
St. James'-wort	Ragwort	*Senecio Jacobæa, L.*
St. John's Wort	St. John's Wort	*Hypericum perforatum, L.*
St. Peter's Wort	Cowslip	*Primula veris, L.*
	Square Stalked St. John's Wort	*Hypericum quadrangulum, L.*
Saltwort	Sea Thrift	*Salsola Kali, L.*
Sandwort	Sandwort	*Genus Arenaria, L.*
Sapwort	Dog's Mercury	*Mercurialis perennis, L.*
Saw-wort	Saw-wort	*Serratula tinctoria, L.*
Scabwort	Elecampane	*Inula helenium, L.*
Scallewort	Clary	*Salvia sclarea, L.*
Scourwort	Soapwort	*Saponaria officinalis, L.*
Sealwort	Common Solomon's Seal	*Polygonatum multiflorum, All.*
Setterwort	Bear's Foot	*Helleborus viridis, L.*
	Green Hellebore	
Sharewort	Sea Aster	*Aster tripolium, L.*
Sicklewort	Self Heal	*Prunella vulgaris, L.*
	Bugle	*Ajuga reptans*
	Herb Carpenter	

*This wort, taken from Mr. Britten's " Plant Names," is given in the hope that some one who may have run across it elsewhere can identify it for me.

Sight-wort	Greater Celandine	*Chelidonium majus, L.*
Skir-wort	Skirret or Water parsnip	*Sium sisarum*
Sleepwort	Lettuce	*Lactuca virosa, L.*
Smallwort	Lesser Celandine	*Ranunculus ficaria, L.*
Smerewort Smearwort	Good King Henry	*Chenopodium Bonus Henricus, L.*
Snapwort	Greater stitchwort	*Stellaria holostea, L.*
Sneezewort Nesewort	Yarrow, Milfoil	*Achillea millefolium, L.* *Achillea Ptarmica, L.*
Soapwort	Scourwort	*Genus Saponaria*
Somerwort	Birthwort	*Genus Aristolochia*
Sowdwort	Columbine	*Aquilegia vulgaris, L.*
Spearwort	Spearwort	*Ranunculus flammula, L.*
Sparewort	Spearwort	*Ranunculus flammula, L.*
Spiderwort	Spiderwort	*Tradescantia virginica, L.*
Spleenwort	Maidenhair fern	*Asplenium trichomanes, L.*
Spoonwort	Common scurvy grass	*Cochlearia officinalis, L.*
Springwort	Any so-called Keyflower	
	Spring flowers	
	Flax	*Linum perenne, L.*
	Cowslip	*Primula veris, L.*
Spurwort	Field Madder	*Sherardia arvensis, L.*
Squinancy-wort	Quincy-wort	*Asperula cynanchica, L.*
Stabwort	Wood sorrel	*Oxalis acetosella, L.*
Staggerwort Stammerwort	Ragwort	*Senecio Jacobæa, L.*
Standelwort	Early Purple Orchid	*Orchis mascula, L.*
Starchwort	Lords and Ladies	*Arum maculatum, L.*
Starwort	Starwort	*Genus Aster* *Genus Stellaria*
Steepwort	Common Butterwort	*Pinguicula vulgaris, L.*
Sticklewort	Agrimony	*Agrimonia Eupatoria, L.*
Stickwort	Wood Sorrel	*Oxalis Acetosella, L.*
Stitchwort	Starwort	*Genus Stellaria*
Strikewort		*Stellaria holostea, L.*
Strapwort	Strapwort	*Corrigiola littoralis, L.*
Stubwort	Wood sorrel	*Oxalis acetosella, L.*
Sulphurwort	Sulphur-weed	*Peucedanum officinale, L.*
Swallow-wort	Greater Celandine	*Chelidonium majus, L.*
Tent wort	Wall rue	*Asplenium Ruta-muraria, L.*
Tetter wort	Greater Celandine	*Chelidonium majus, L.*
Thimble-wort	Foxglove	*Digitalis purpurea, L.*
Throatwort	Knotted Figwort	*Scrophularia nodosa, L.* *Various campanulas*
Thrumwort	Thrumwort	*Amaranthus caudatus, L.*
Toothwort	Shepherd's purse	*Capsella Bursa-pastoris, Medik.* *Genus Dentaria*
Toywort	Shepherd's purse	*Capsella Bursa-pastoris, Medik.*
Wallwort	Dwarf elderberry	*Sambucus ebulus, L.*
	Wall pepper	*Sedum, acre, L.*
	Pellitory-of-the-wall	*Parietaria officinalis, L.*
	Pennywort	*Cotyledon umbilicus, L.*
	Navelwort	*Cotyledon umbilicus, L.*
Walewort or	Dwarf Elder	*Sambucus ebulus, L.*
Walwort	Dwarf Elder	*Sambucus ebulus, L.*
Wartwort	Sun Spurge	*Euphorbia helioscopia, L.*
	Turnsole	
	Greater Celandine	*Chelidonium majus L.*
Waterwort	Maidenhair	*Asplenium Trichomanes, L.* *Genus Elatine*
Waywort	Scarlet pimpernel	*Anagallis arvensis, L.*
Whitewort	Feverfew	*Pyrethrum parthenium, L.*
	Whorled Solomon's Seal	*Polygonatum multiflorum, Moench.*
Wildwort	Lungwort—Our Lady's Milkwort	*Pulmonaria officinalis, L.*
Willow-wort	Great yellow loosestrife	*Lysimachia vulgaris, L.*
Wolfs-wort	Monkshood	*Aconitum Napellus, L.*
Woundwort	Kidney Vetch	*Anthyllis vulneraria, L.*
	Golden Rod	*Solidago virgaurea, L.* *Species Stachys*
Wreathe-wort	Purple orchis	*Orchis mascula, L.*
Yellow-wort	Yellow Centaury	*Chlora perfoliata, L.*
Youth-wort	Sundew	*Drosera rotundifolia, L.*

The Partridge Fresco. From Pavilion of "Caravanserai," Frontispiece of "Palace of Minos." Vol. II, Part 1, Sir Arthur Evans. By permission of the Macmillan Co., Publishers.

"RIGHTE DITTANY"

HARRIET ADDAMS BROWN

ONE of the most interesting phases of herb growing is the struggle to secure for our modern gardens authentic species of the herbs mentioned in the old records. These may be found listed under many different names and wander from genus to genus to the complete mystification of the unwary.

"Dictamnus" or "Dittany" is one of the oldest of these medical herbs of mystery. Headed by Hippocrates himself, and including among the more familiar names Apuleius, Theophrastus, Dioscorides, Pliny, Plutarch, Cicero and Virgil, the long list of ancient sages who discussed the provenance and virtues of this Cretan Herb, reads like the catalogue of a classical library of the Golden Age.

Theophrastus (Sir A. Hort's translation) has perhaps the most comprehensive description. "Dittany is peculiar to Crete. This plant is marvellous in virtue and is useful for many purposes. . . . Its leaf is like Penny-royal, to which it also bears some resemblance in taste, but the twigs are slenderer. They use the leaves, not the twigs nor the fruit. . . . The leaf is useful . . . above all against difficult labour in women. . . . It is given as a draught in water. It is a scarce plant, for the region which bears it is not extensive, and the goats graze it down because they are fond of it. The story of the arrows is said to be true—that if goats eat it when they have been shot it rids them of the arrow. This, then, is Dittany and such its properties. The virtue of Dittany is perceived directly it is taken into the mouth, for a small piece has a warming effect."

Virgil's description is the most vivid in the famous passage which tells of Venus helping to heal the wound of Æneas:

> "A branch of healing Dittany she brought
> Which in the Cretan fields with care she sought.
> Rough is the stem, which woolly leaves surround.
> The leaves with flowers, the flowers with purple crowned.
> Well known to goats; a sure relief
> To draw the pointed steel and ease the grief."
>
> *Æneid, Bk. XII.* (DRYDEN)

The classical belief in the medicinal properties of Dittany continued for generations, but while the old writers seem to have agreed as to the appearance and properties of the plant, by the first quarter of the 16th century a note of botanical confusion creeps into the records.

Caspar Bauhin, in his *Theatri Botanici*, Basle, 1523, at the end of a long list of Dictamnums, adds, "*D. Albus Vulgo, Sive Fraxinella.*"

In 1548 William Turner writes: "Dictamnus groweth in Cady and hath rounde, thick, rough leaves. I have not seen it growing but dry oft. It may be had in Venice and in Antwerp. Some abuse for this 'Fracinella.' Dictamnus may be named in English 'Righte Dittany.'" Twenty years later he triumphantly writes: "I have seen Dittany growing in England in Maister Riches garden naturally, but it groweth nowhere else that I know saving only in Candy."

He also states that "Lepidum is called in English 'Dittani,' but foolishly and unlearnedly. In Duche 'Pepperkrout,' that is, 'Pepperwort,' because it is so exceedynge hote, which name were more fitter in English for this herb than the name of 'Dittani.' That the name

Photograph of Origanum Dictamnus L. from the Gray Herbarium of Harvard University, courtesy of Mr. C. A. Wetherby, Curator.

of 'Dittani' might abyde proper unto ye 'Righte Dittano' which begynneth now to be set and sowen in England."

The *Hortus Kewensis* confirms Turner, and lists "Dittany of Crete" as "cultivated in 1551 by Mr. John Riche." The *Hortus* also records it as "Cretan Marjoram," "cultivated by Mr. John Gerard."

In 1764 Linneus, in his *Genera Plantarum*, referring to Bauhin, removed Dittany from the genus "Dictamnum" to the genus "Origanum," and we may find our herb from that time on listed as *Origanum Dictamnus, L.*

Thanks, however, to the researches of the distinguished archæologist, Sir Arthur Evans, whose excavations in Crete have turned fairy tale to fact, "Righte Dittany" has received still further proof of identity and we can stand firm against the usurping claims made by the bastard dittany, *Dictamnus Fraxinella*, to the heritage of centuries of lore and legend which rightfully belongs to *Origanum Dictamnus*.

Sir Arthur Evans has unearthed a fresco in Crete, near the ancient town of Knossos. Once it decorated the walls of what he has named "The Caravanserai," an ancient rest house of the middle Minoan period, that unbelievable era of civilization of 1500 B.C. It is a frieze of birds, perched on bushes and plants. One of these plant forms so closely resembles *Dictamnus* as described by the ancient writers, and is so similar to the plant now growing on the rocky Cretan cliffs, that we may feel quite confident that here is what must be the earliest extant picture of this "literary celebrity" among herbs, famous down the ages as the "Marvellous healing herb of Crete," and still called "Diktamno" by the Cretan peasants. Sir Arthur describes the living plant which he found in the crevices of the rocks as having "Round leaves like Penny-royal . . . lilac-veined and covered with soft, downy hairs, answering Virgil's description. The flower which emerges from a cluster of overlapping bracts is labiate, and is also of a delicate purple hue." [1]

[1] From *Palace of Minos at Knossos*, Vol. II, Part 1. By permission of the Macmillan Co., Publishers.

Both this description and the fresco itself are corroborated by the unique specimen of *Origanum Dictamnus* now in the Gray Herbarium of Harvard University, which was collected in 1846 by Theodor Von Heldreich, Director of the Botanic Garden at Athens. Except for comparative scale, which was an unimportant point with early decorators, the shape of the leaves and the habit of growth of the painted plant leave little doubt but that in these prehistoric fragments we have a vindication of the identity of Dittany and a record which gives us an unbroken plant pedigree for over three thousand years.

The restoration of this historic plant to common knowledge and its propagation on these alien shores would be an interesting and worth-while project for those members of the Herb Society who "delight in looking backward."

Bibliography

A. Musa: *Apuleius Barbarus.*

Meursius: *Creta,* Lib. II, p. 109.

Caspar Bauhin: *Theatri Botanici,* Basle, 1523.

William Turner: *The Names of Herbs,* 1548.
Herball, 1568.

William Coles: *Adam in Eden,* 1657.

Gerard: *Herball.*

Joannes Raius: *Plantarum Historiam.*

John Quincy, M.D.: *A Compleat English Dispensatory,* 1749.

Woodville: *Medical Botany Supplement,* 1779.

Kniphof: *Botanica in Originali,* 1758.

Zorn: *Icones Pl. Med.,* 1779.

Curtis: *Botanical Magazine,* 1795, Vol. IX.

Sir Arthur Evans: *Palace of Minos,* Vol. II, Part 1.

R E C I P E S

HERB VINEGARS

BEFORE the next bulletin of the H. S. A. enters press, the fragrance of the herb garden will be ours again, and we shall be garnering the various potencies of our herbs for use when snow covers the ground.

That wine or vinegar has ever been a favorite medium to preserve the fresh flavors of sweet and savory herbs is recognized in the very earliest cook books, but a good bit of epicurean delight is still left for us today in trying out and using these quaint old rules in our modern kitchens.

We may not exhibit on the pantry shelf the 65 sorts of variously flavored vinegars which made a Paris shop famous in 1800—I have always hoped to find somewhere the list of flowers and leaves which that *vinaigrier* steeped, and which, it is recorded, were indispensable in the French cuisine or pharmacopœia.

Besides sweet herbs, spicy, aromatic flowers were made into "flower vinegars"—roses, violets, clovegilly or spice pink and elder blows.

Rosemary vinegar was most easily made by steeping the flowers and fresh, leafy tips, and that favorite decoction seems to have been as useful for boiled beef as aching heads.

Burnet and Basil are two herbs which lose much of their aromatic oils in drying, so the flavors were preserved in vinegar. They are leafy, succulent herbs in the summer gardens, of great use in tomato and fish cookery and salads.

The Cook's Oracle—a revision published in 1823—reminds us that "the flavor of Burnet leaves resembles cucumber so exactly that when infused in vinegar, the nicest palate would pronounce it to be cucumber."

Basil vinegar, the *Oracle* goes on to say, "is a secret the makers of Mock Turtle may thank us for telling—a tablespoon of this put in when the Soup is finished will impregnate a Tureen of Soup with Basil and acid flavors at very small cost when fresh Basil and Lemons are extravagantly dear!"

Vinegars of Tarragon, Mint, Basil and Burnet, Chives, Balm and Savory are best made by pounding the fresh leaves down into a stone crock or bottle, covering with vinegar, sealing tightly and leaving ten days or more if desired. Strain, and renew the herbs if a stronger potion is desired, when it may be filtered and stored in smaller containers.

In Parmentier's *Art de Faire les Vinaigres* there is a practical carefully worded rule for a useful, all-round salad vinegar.

> "Take of Tarragon, Savory, Chieves, Eshallots, 3 ounces each, a handful of the tops of Mint and Balm all dry and pounded; put into a wide mouthed Bottle with a gallon of the best Vinegar; cork it close; set it in the Sun; and in a fortnight, strain off and squeeze the Herbs; let it stand a day to settle and then strain it through a filtering bag."

I quote this recipe exactly, though from other experiences of this author it would seem that the drying of the herbs first was a mistake.

I asked a friend if she had Tarragon in her garden. "Yes," she answered, "but we've eaten it all up!" So as we use the fresh succulence of our herbs all through the summer, let's plant more and more of every kind, that from our tables some may be spared for the age-old herb and flower vinegars.

—Helen Noyes Webster

SORREL SOUP (for 6)

Wash Sorrel and put in saucepan with a little water (not covered). Cook slowly for about ½ hour. Put 4 cups of milk with small white onion (whole) in double boiler. Add 2 tablespoonfuls of butter and 2 tablespoonfuls of flour to the hot milk. Let stand, and add Sorrel and strain. Season.

Use about a handful of the Sorrel.

A Refreshing Drink for Hot Days

One quart strong Orange Pekoe Tea. Boil a double handful of fresh Balm leaves in a quart of water. Strain and add to tea. When cold, add 1 quart ginger ale. Ice, and serve with sprigs of fresh Balm.

An Excellent Stuffing for Tomatoes, Onions and Sweet Green Peppers

2 shallots
1 teaspoon parsley
¼ teaspoon thyme leaves
¼ teaspoon winter savory
¼ teaspoon basil

Put through meat chopper before measuring
½ cup chopped ham
½ cup bread crumbs
2 oz. melted butter

OMELETTE AUX FINES HERBES

For each egg allow:
1 tablespoonful of water
Scant ¼ teaspoonful flour
Salt, pepper
1 teaspoonful of fresh omelet herbs, finely chopped, or a generous pinch of dried omelet herbs.

For each omelet allow: ¾ to 2 tablespoonfuls of butter, depending on the size of the pan. The butter must come far enough up the sides of the pan to prevent the edges of the omelet from sticking.

Place water and flour in a bowl with salt and pepper and mix till smooth. Then break in the eggs, and beat with a tablespoon until whites and yolks are thoroughly blended. Take care not to mix in air during the beating. The herbs may be added to the water or to the omelet after blending, as preferred.

A good herb mixture may be made from fresh Chives, Marjoram and Thyme, with or without Basil. However, single herbs or other combinations are also satisfactory. Chopped parsley, reënforced with onion juice or dried celery tips, makes a good winter combination.

While mixing the omelet, place the butter in a smooth frying pan, plenty large enough to insure a thin omelet. A 7- or 8-inch pan is about right for a one-egg omelet, and it is better to make several individual omelets than a too-thick omelet of family size. Let the butter bubble slowly over a low flame until the water has been driven off and the butter barely begins to brown, then increase the heat and pour in the omelet mixture, lifting it at the edges so that the uncooked mixture will run under. It takes from 2 to 3 minutes to cook and brown a thin omelet.

Roll the omelet toward the handle of the pan, turn it around and slip on a hot plate or platter to be served immediately.

All tinned vegetables are improved by adding a pinch of herbs when boiling.

* * *

Scour the table top before meals with a large handful of Balm.

* * *

A garnish for all green salads: *Finely-cut sage mixed with cream cheese and formed into balls.*

FRESH DILL

Finely-chopped leaves of fresh dill are delicious in chicken gravy and in cream sauces used with chicken or fish. Use enough of the dill to color the sauce quite a bright green. Chopped fresh dill may be sprinkled on broiled chops or steaks, covered with hot melted butter, and allowed to stand for a few minutes in a hot oven before serving.

Now that the herb garden has again come into its own it may be interesting to consider some of the designs used in the early 18th century as a possible motif for our present-day gardens. "The Complete English Gardner," written in 1704 by Leonard Meager, presents "Divers forms or plots for gardens," a few of which are shown here.

SOME HERBS FROM THE OLD WORLD AND THE NEW
THAT WILL GROW IN NEW ENGLAND

ALICE T. WHITNEY

"From Greenland's icy mountains,
From India's coral strand."
and from elsewhere—far and near, North and South, East and West.

* * *

". . . and such gardens are not made
By singing:—'Oh, how beautiful!' and sitting in the shade."
(From: *The Glory of the Garden*—KIPLING)

THE list of herbs following includes everything that grows at our Milton place, whether it had been there for a century or more, or for a week or less, when the ground froze up for the winter this past December. It includes things whether there for years; whether got for ornament only; whether got because they were native to this region—or whether got just because they *are* "herbs."

The authorities referred to as to "what *is* an herb" are:

A Modern Herbal by Mrs. M. Grieve and Mrs. C. F. Leyel, 1931.

A Synopsis of Medical Botany of the U. S. by J. M. G. Carter, M.A., M.D., Ph.D., Sc.D., 1888.

And a similar book by Charles A. Lee, M.D., *A Catalogue of the Medicinal Plants, Indigenous and Exotic, Growing in the State of New-York*, 1848.

A Manual of Botany by Laurence Johnson, A.M., M.D. (from a series, "Wood's Library of Medical Authors"), 1884.

Missouri Botanical Garden Bulletin, Vol. 10, No. 4, April, 1922.
Professors Carter and Miller are very spacious in their opinions as to what they call "Medicinal Plants," which generous latitude has caused this list to become so incredibly long, but they really did write as if nearly "*Anything* green that grew out of the mould (in these U. S.) was an excellent herb to our fathers of old."

Other books have been consulted, but any information as to what is what that others may give is included in those above.

For botanical information the following books have been continually referred to:

Bailey's *Encyclopædia of Horticulture*, 1922.
Bailey's *Hortus*, 1930.
Gray's *New Manual of Botany* (Seventh Edition), 1908.
A Manual of Weeds by Ada E. Georgia, 1931.
Standardized Plant Names, 1924.
The Flower Finder by George L. Walton, M.D., 1930.

Also many others in the past, but only those above just now while hastily compiling these lists.

Liatris spicata (and other sp.)	Spike Gayfeather (Button Snakeroot)
Lobelia cardinalis	Cardinal Flower
Lobelia siphilitica	Great Lobelia (Great Blue Lobelia)
Lycopodium clavatum (and other sp.)	Common Club Moss
Medeola virginiana	Indian Cucumber Root
Mentha aquatica	Wild Water Mint
Mentha arvensis var. canadensis (M. canadensis)	Canada Mint (American Wild Mint)
Menyanthes trifoliata	Bogbean (Buck Bean, etc.)
Monarda fistulosa	Wild Bergamot
Monarda punctata	Horsemint
Oxalis violacea	Violet Woodsorrel
Osmorhiza longistylis	American Sweet Cicely
Panax quinquefolium	Dwarf Gensing (Gound Nut)
Panax trifolium	Gensing
Pedicularis canadensis	Early Wood Betony (Lousewort)
Physostegia virginiana	False Dragonhead
Phytolacca decandra	Pokeweed
Podophyllum peltatum	Mayapple (American Mandrake)
Polemonium reptans	Greek Valerian (Abscess Root)
Polygonatum biflorum	Small Solomon's Seal
Polygonatum commutatum	Great Solomon's Seal
Potentilla Anserina	Silverweed
Pyrola americana	Roundleaf Pyrola
Sagittaria latifolia	Arrowhead (Wampee) etc.
Sanguinaria canadensis	Bloodroot
Sanguisorba canadensis	American Burnet (Canadian Burnet)
Satureia vulgaris (Clinopodium vulgaris)	Wild Basil (Hedge Basil) etc., etc.
Scutellaria galericulata	Common Skullcap
Silene virginica (and other sp.)	Fire Pink (Catch Fly)
Silphium laciniatum	American Ipecac (Rosin Weed)
Silphium trifoliatum	Cup Plant (Rosin Weed)
Solidago odora	Fragrant Goldenrod
Stylophorum diphyllum	Celandine Poppy
Thalictrum dioicum	Early Meadow-Rue
Tiarella cordifolia	Foamflower (Coolwort) (False Mitrewort)
Tradescantia bracteata	Spiderwort (Pink)
Tradescantia virginiana	Spiderwort (Common)
Trillium erectum	Purple Trillium (Beth Root)
Trillium nivale	Dwarf Trillium
Trillium grandiflorum	Snow Trillium
Triosteum perfoliatum	Horse Gentian
Verbena hastata	Blue Vervain
Vernonia noveboracensis	Tall Ironweed
Veronica Virginica (Leptandra Va.)	Culver's Root (Black Root) etc.
Viola palmata	Wild Okra
Viola striata (and many other sp.)	White Violet

Herbs not native to the U. S., at Winter Valley, Milton, Massachusetts. (As of December, 1934.)

Alchemilla vulgaris	Ladies' Mantle
A. alpina	Ladies' Mantle, Alpine
Allium Cepa	Onion
Ajuga reptans	Bugle
Anemone Pulsatilla (and other sp.)	European Pasqueflower (Pulsatilla)
Armeria maritima (Statice Armeria)	Common Thrift
Arnica montana	Arnica
Artemisia mutellina	———
Asparagus officinalis	Asparagus
Asperula odorata	Sweet Woodruff
Linum alpinum	Alpine Flax
L. austriacum	Austrian Flax
Lupinus polyphyllus	Lupine (Washington Lupine)
Lysimachia Nummularia	Moneywort
L. vulgaris	Golden Loosestrife (Yellow Loosestrife)
Lythrum salicaria	Purple Loosestrife
Mentha requieni (Thymuscorsica)	Corsican Mint (Requiem Mint) (Grows like a weed here and much of it survives worst winters)

Liatris spicata (and other sp.)	Spike Gayfeather (Button Snakeroot)
Lobelia cardinalis	Cardinal Flower
Lobelia siphilitica	Great Lobelia (Great Blue Lobelia)
Lycopodium clavatum (and other sp.)	Common Club Moss
Medeola virginiana	Indian Cucumber Root
Mentha aquatica	Wild Water Mint
Mentha arvensis var. canadensis (M. canadensis)	Canada Mint (American Wild Mint)
Menyanthes trifoliata	Bogbean (Buck Bean, etc.)
Monarda fistulosa	Wild Bergamot
Monarda punctata	Horsemint
Oxalis violacea	Violet Woodsorrel
Osmorhiza longistylis	American Sweet Cicely
Panax quinquefolium	Dwarf Gensing (Gound Nut)
Panax trifolium	Gensing
Pedicularis canadensis	Early Wood Betony (Lousewort)
Physostegia virginiana	False Dragonhead
Phytolacca decandra	Pokeweed
Podophyllum peltatum	Mayapple (American Mandrake)
Polemonium reptans	Greek Valerian (Abscess Root)
Polygonatum biflorum	Small Solomon's Seal
Polygonatum commutatum	Great Solomon's Seal
Potentilla Anserina	Silverweed
Pyrola americana	Roundleaf Pyrola
Sagittaria latifolia	Arrowhead (Wampee) etc.
Sanguinaria canadensis	Bloodroot
Sanguisorba canadensis	American Burnet (Canadian Burnet)
Satureia vulgaris (Clinopodium vulgaris)	Wild Basil (Hedge Basil) etc., etc.
Scutellaria galericulata	Common Skullcap
Silene virginica (and other sp.)	Fire Pink (Catch Fly)
Silphium laciniatum	American Ipecac (Rosin Weed)
Silphium trifoliatum	Cup Plant (Rosin Weed)
Solidago odora	Fragrant Goldenrod
Stylophorum diphyllum	Celandine Poppy
Thalictrum dioicum	Early Meadow-Rue
Tiarella cordifolia	Foamflower (Coolwort) (False Mitrewort)
Tradescantia bracteata	Spiderwort (Pink)
Tradescantia virginiana	Spiderwort (Common)
Trillium erectum	Purple Trillium (Beth Root)
Trillium nivale	Dwarf Trillium
Trillium grandiflorum	Snow Trillium
Triosteum perfoliatum	Horse Gentian
Verbena hastata	Blue Vervain
Vernonia noveboracensis	Tall Ironweed
Veronica Virginica (Leptandra Va.)	Culver's Root (Black Root) etc.
Viola palmata	Wild Okra
Viola striata (and many other sp.)	White Violet

Herbs not native to the U. S., at Winter Valley, Milton, Massachusetts. (As of December, 1934.)

Alchemilla vulgaris	Ladies' Mantle
A. alpina	Ladies' Mantle, Alpine
Allium Cepa	Onion
Ajuga reptans	Bugle
Anemone Pulsatilla (and other sp.)	European Pasqueflower (Pulsatilla)
Armeria maritima (Statice Armeria)	Common Thrift
Arnica montana	Arnica
Artemisia mutellina	———
Asparagus officinalis	Asparagus
Asperula odorata	Sweet Woodruff
Linum alpinum	Alpine Flax
L. austriacum	Austrian Flax
Lupinus polyphyllus	Lupine (Washington Lupine)
Lysimachia Nummularia	Moneywort
L. vulgaris	Golden Loosestrife (Yellow Loosestrife)
Lythrum salicaria	Purple Loosestrife
Mentha requieni (Thymuscorsica)	Corsican Mint (Requiem Mint) (Grows like a weed here and much of it survives worst winters.)

Bellis perennis..............English Daisy

Chenopodium bonus henricus..Good King Henry("Mercury")etc.
Cimicifuga fœtida simplex....Kamchatka Bugbane
(*C. racemosa simplex*)
Clematis recta..............Ground Clematis
Colchium autumnale.........Meadow Saffron
Convallaria majalis..........Lily-of-the-Valley
(and 2 vars.)
Convallaria majalis rosea....Pink Lily-of-the-Valley
Crocus sativus..............Saffron Crocus

Dictamnus Fraxinella........Dittany
(*D. albus*)

Ephedra sp.................Ephedra (Jointfir) (Mah Huang)

Fumaria officinalis..........Common Fumatory

Galanthus nivalis............Snowdrop
(and other sp.)
Gossypium hirsutum.........Upland Cotton

Helianthemum mutabile......Sunrose
(and other sp.)
Heliotropium peruvianum.....Heliotrope
(Hort. vars.)
Helleborus niger.............Black Hellebore (Christmas Rose)
Hemerocallis fulva...........Orange Day Lily
Herniaria glabra............Burstwort
H. hirsuta...............Burstwort
Hesperis matronalis..........Rocket (Sweet Rocket) (Dames' Rocket)
Hieracium auranticum.......Orange Hawkweed ("Grim the Collier," etc.)

Iris fiorentina..............Orris
I. germanica.............Orris
I. pallida...............Orris
I. pseudoacorus...........Yellowflag Iris

Lamium album.............White Dead Nettle
L. maculatum............Dead Nettle
Linaria origanifolia.........——— (A delightful creeping plant
(and other sp.) in shade. Well worth while.)
Linum usitatissimum........Flax (*The* flax of commerce)

Myosotis scorpiodes..........Forget-me-not (True)
(*M. palustris*) (and other sp.)
Muscari botryoides..........Grape Hyacinth
(and other sp.)

Narcissus Jonquilla.........Jonquil
(and hybrids)
N. poeticus...............Poets' Narcissus
N. Pseudo-narcissus.......Daffodil (Trumpet) (Common Daffodil)
(Many vars.)

Pæonia officinalis...........Peony
(and Hort. vars.)
Polygonum bistorta..........Bistort
Primula elatior............Oxlip
P. veris.................Cowslip
P. vulgaris..............Primrose, Common (English Yellow)
Pulmonaria angustifolia
cærulea (or: *P. a. azurea*) ..Lungwort (Blue) (A most delightful, easy and indispensable spring plant—very hardy)
P. maculata..............Lungwort, Common

Radicula Armorica ....Horseradish
(*Cochlearia Armorica*)
(and now: *Armorica rusticana*)
Radicula Nasturtium-aquatica.Watercress

Sanguisorba officinalis........Burnet
Satureia croatica...........——— Can find no English name but it seems as if it should be called "Croatian Savory." A delightful, creeping plant, very much like *Linaria origanifolia*
Scilla mutans (*Scilla nonscripta*)
(*Hyacinthus nonscriptus*)...English Bluebell

Taraxacum officinale........Dandelion (This is deliberately grown in a row in vegetable garden for spring salad and, mysteriously, seems not to add to the usual weed dandelions on the place!)
Thymus belgradensis
T. chamædrys
T. lanicaulis
(also all the sp. we grow at Woods Hole)
Tusilago Farfara...........Coltsfoot

Valeriana officinalis	Valerian
Vinca minor (Blue, White and "Red") (Also "Bowles Variety," an extra good and robust blue.)	Periwinkle (Myrtle) (Joy of the Ground)
Viola odorata (Several vars.)	Sweet Violet
V. tricolor	Heartsease (Johnny-Jump-Ups) etc., etc.

Herb plants and shrubs found growing wild at Winter Valley, Milton, Massachusetts. Nearly all are the most arrant weeds, some are villainous and a *few* are attractive and desirable.

N=NATIVE U. S.

N	*Achillea Millefolium*	Yarrow (Milfoil)
N	*Acorus Calamus*	Sweet Flag (Sweet Sedge)
N	*Agrimonia mollis*	Soft Agrimony
	Agropyron repens	Quack-Grass (Couch-Grass) etc.
N	*Alisma Plantago-aquatica*	Water Plantain
N	*Ambrosia artemisiifolia* (ALAS!)	Common Ragweed (sic) (That enemy of the human race!)
N	*Abrosia trifida*	Great Ragweed (Giant Ragweed)
N	*Anaphalis margaritacea*	Pearl Everlasting, etc.
N	*Anemone quinquefolia*	Wood Anemone
N	*Apocynum androsæmifolium*	Spreading Dogbane
N	*Apios tuberosa*	Wild Bean (Ground Nut)
N	*Arisæma triphyllum*	Jack-in-the-Pulpit
N	*Asclepias syriaca*	Common Milkweed
N	*Aster cordifolius*	Blue Wood Aster
N	*Aster novæ-angliæ* (and other sp.)	New England Aster
	Barbarea verna (*B. præcox*)	Early Winter Cress (Scurvy Grass)
N	*Barbarea vulgaris*	Winter Cress (St. Barbara's Cress, etc., etc.)
	Brassica nigra	Black Mustard
N	*Caltha palustris*	Marsh Marigold
	Capsella Bursa-pastoris	Shepherds' Purse
	Cerastium vulgatum	Mouse-Ear Chickweed
	Chelidonium majus	Greater Celandine
	Chenopodium album	Smooth Pigweed (Lambs' Quarters, etc.)
	Chrysanthemum Leucanthemum	Common Daisy (Ox-Eye Daisy, etc.)
	Cichorium Intybus	Chicory
	Daucus Carota	Wild Carrot (Queen Anne's Lace, etc.)
	Nepeta Glechoma (*N. hederacea*)	Ground Ivy, etc., etc.
N	*Oenothera biennis* (and other sp.)	Evening Primrose
N	*Oxalis stricta*	Wood Sorrel
	Plantago lanceolata	English Plantain (Ribwort) (a curse anywhere)
	Plantago major	Common Plantain
N	*Polygonum Hydropiper*	Smartweed
N	*Polygonum hydropiperoides*	Mild Water Pepper
	Polygonum Persicaria	Lady's Thumb, etc.
	Portulaca oleracea	Purslane, etc., etc.
N	*Potentilla argentea*	Silvery Cinquefoil
N	*Potentilla canadensis*	Common Cinquefoil
N	*Potentilla monspeliensis*	Rough Cinquefoil (Norway Cinquefoil)
N	*Prenanthes alba* (*Nabalus alba*)	White Lettuce (Rattlesnake Master, etc.)
	Prunella vulgaris	Selfheal (Prunella) (Carpenter's Herb)
	Ranunculus acris	Tall Buttercup
N	*Ranunculus repens*	Creeping Buttercup
	Rubus sp.	Wild Blackberry
	Rumex Acetosella	Field Sorrel (Sheep Sorrel)
N	*Senecio aureus*	Golden Groundsel (Life Root) (Squaw Root)
N	*Smilacina racemosa*	False Solomon's Seal
N	*Solidago sp.*	Goldenrod (any and all are "Herbs," apparently)
N	*Spiræa latifolia*	Meadow-sweet
N	*Spiræa tomentosa*	Hardhack (Steeplebush)
	Stellaria media	Common Chickweed
N	*Symplocarpus fœtidus*	Skunk Cabbage

N *Equisetum arvense*.........Common Horsetail (Field Horsetail)
N *Erigeron annuus*..........Daisy Fleabane
N *Erigeron pulchellus*........Robin's Plantain (Blue Spring Daisy, etc.)
N *Eupatorium perfoliatum*....Boneset (American) (Thoroughwort, etc.)
N *Eupatorium purpureum*....Joe-Pye Weed (Gravelroot, etc.)

Fragaria (Sp. unknown) Wild Strawberry

N *Galium asprellum*..........Rough Bedstraw
N *Geranium maculatum*......Wild Geranium (American Cranesbill Root)
N *Gnaphalium polycephalum*..Common Everlasting (Old Field Balsam), etc.
N *Gnaphalium uliginosum*....Low Cudweed

N *Hypericum mutilum*........Dwarf St. Johnswort
Hypericum perforatum.....Common St. Johnswort
N *Hypoxis hirsuta* (*H. erecta*). Yellow Star-Grass(Goldeye-Grass)

N *Impatiens fulva* (*I. biflora*) .Spotted Jewelweed (Spotted Touch-Me-Not)
N *Iris versicolor*.............Blue Flag Iris (Larger Blue Flag)

N *Lactuca canadensis*........Wild Lettuce
Leontodon autumnalis......Fall Dandelion
Linaria vulgaris...........Common Toadflax (Butter and Eggs)
N *Lobelia inflata*...........Indian Tobacco
N *Lycopus virginicus*.........Bugleweed (Water-Horehound) etc.
N *Lysimachia quadrifolia*.....Four-Leaved Loosestrife (and other sp.)

Malva rotundifolia.........Dwarf Mallow (Cheeses, etc.)
N *Mollugo verticillata*........Indian Chickweed (Carpetweed) (A dreadful pest)
N *Monotropa Hypopitys*......Pinesap (False Beechdrops)

Tanacetum vulgare.........Tansy
Taraxacum officinale.......Dandelion
N *Typha latifolia*............Cat-tail (Common)

Urtica dioica.............Stinging Nettle (Great Nettle)
N *Uvularia sessilifolia*.......Little Merribells (Bellwort) (*Oakesia sessilifolia*)

Verbascum Thapsus.......Common Mullein
N *Veronica officinalis*........Common Speedwell (The most charming and useful of weeds. Makes a fine, nearly-evergreen ground cover in shade where *nothing* else will grow.)
N *Veronica serpyllifolia*......Thyme-leaved Speedwell (Even this is useful, in shade-leaves so much like V. repens that have often to stop and look to make sure whether or not it is the latter—advise everyone not to scorn either of these two friendly and pleasant weeds.)
N *Viola blanda*..............Sweet White Violet
N *Viola cucullata*............Blue Marsh Violet
N *Viola papilionacea*.........Butterfly Violet
Vicia sativa..............Common Vetch (Tare)

FERNS

N *Onoclea sensibilis*........Sensitive Fern
N *Osmunda cinnamomea*....Cinnamon Fern
N *Osmunda regalis*.........Royal Fern

N *Parthenocissus quinquefolia*) Virginia Creeper (Woodbine, etc.) (*Psedera quinquefolia*) (*Ampelopsis quinquefolia*)

SHRUBS

Rosa sp................Wild Rose?
N *Rhus glabra*............Smooth Sumach
N *Rhus Toxicodendron*(Alas)Poison Ivy
N *Rhus typhina*...........Staghorn Sumach
N *Sambucus canadensis*....Elder (American)
N *Smilax rotundifolia*......Greenbrier (Horse Brier, etc.)
N *Viburnum dentatum*.....Arrow-Wood
N *Vitis sp.*...............Wild Grapes

Tree, shrub and vine "herbs" at Winter Valley, Milton, Massachusetts. (As of December, 1934.) ***All native to the U. S.***

Abies balsamea..............Balsam Fir
Abies Frazeri...............Frazer Fir (and other sp.)
Acer negundo...............Box Elder
Acer rubrum...............Red Maple
Acer saccharinum..........Silver Maple (White Maple)
Acer saccharum............Sugar Maple
Amelanchier canadensis......Shadbush (Downy Shadblow) etc.

Amelanchier lævis.......... Allegheny Shadblow
Amelanchier stolonifera...... Running Shadblow
Aralia spinosa.............. Hercules Club (Angelica Tree)
Arctostaphyllos uva-ursi...... Bearberry

Benzoin æstivale............. Spicebush (Benjamin Bush) etc.
(*Lindera Benzoin*)
Betula lenta................ Sweet Birch
Betula lutea................ Yellow Birch (Gray Birch)
Betula papyrifera........... Paper Birch (Canoe Birch)
Betula populifolia........... Gray Birch

Calycanthus floridus......... Sweetshrub (Strawberry Shrub) (Carolina Allspice)
Carpinus caroliniana......... American Hornbeam
(*C. americana*)
Cassia marilandica.......... American Senna
Ceanothus americanus........ New Jersey Tea (Red Root)
Celastrus scandens.......... American Bittersweet (Wax-work)
Celtis occidentalis........... Hackberry (Sugarberry)
Cephalanthus occidentalis..... Buttonbush
Cercis canadensis............ American Judas Tree (American Redbud)
Chiogenes hispidula.......... Creeping Snowberry
Chionanthus virginiana...... Fringe Tree
Cladrastis lutea............. Yellow-wood
(*Virgilia tinctoria*)
Clematis virginiana.......... Virgin's Bower
Clethra alnifolia............. Summersweet (Sweet Pepperbush)
Comptonia asplenifolia....... Sweet Fern
(*Myrica asplenifolia*)
Cornus Amomum............ Silky Dogwood (Kinnikinnik)
(*C. seriaca*)
Cornus circinata (*C. rugosa*).. Roundleaf Dogwood
Cornus florida.............. Flowering Dogwood
Corema Conradi............. Broom-Crowberry
Corylus americana........... Hazelnut (American)
Cratægus coccinea........... Thicket Hawthorn
Cratægus phenopyrum....... Washington Thorn
(*C. cordata*)

Diervilla rivularis........... Georgia Bush-honeysuckle
Diervilla trifida............ Dwarf Bush-honeysuckle
Diospyros virginiana......... Persimmon (Common)
Dirca palustris............. Leatherwood

Empetrum nigrum........... Crowberry

Larix laricina............... American Larch (Tamarack) (Hackmatack)
Ledum grœnlandicum........ Labrador Tea
(*L. latifolium*)
Linnæa borealis americana.... Twinflower
Liquidamber Styraciflua...... Sweet Gum (Bilsted)
Liriodendron tulipifera....... Tulip Tree
Loiseleuria procumbens....... Mountain Azalea

Magnolia acuminata......... Cucumber Magnolia
Magnolia glauca............ Swamp Magnolia (Sweet Bay)
Mahonia Aquifolium......... Oregon Hollygrape
(*Berberis Aquifolium*)
Malus coronaria............. Wild Sweet Crab
(and other sp.)
Mitchella repens............. Partridgeberry (Squaw Vine)
Myrica carolinensis.......... Bayberry
Myrica Gale................ Sweet Gale

Nemopanthus mucronulatus... Mountain Holly
Nyssa sylvatica.............. Tupelo (Sour Gum) (Pepperridge)

Oxydendrum arboreum....... Sorrel Tree (Sour Wood)

Physocarpus opulifolius...... Ninebark
Pieris mariana.............. Staggerbush
(*Lyonia mariana*)
(*Andromeda mariana*)
Pinus Banksiana............ Jack Pine
Pinus resinosa.............. Red Pine
Pinus rigida................ Pitch Pine
Pinus Strobus.............. White Pine
Populus balsamifera......... Balsam Poplar (Old Field Balsam) (Tacmahac)
Populus deltoides............ Cottonwood (Necklace Poplar)
(*P. canadensis*)
Populus tremuloides......... Quaking Aspen
Prunus serotina............. Wild Black Cherry
Prunus virginiana........... Chokecherry
Ptelea trifoliata............. Hop Tree (Wafer Ash)

Quercus alba................ White Oak
Quercus bicolor.............. Swamp White Oak
Quercus coccinea............ Scarlet Oak
Quercus palustris............ Pin Oak
Quercus rubra.............. Red Oak

Botanical name	Common name
Epigæa repens	Trailing Arbutus (Mayflower)
Euonymus americanus	Strawberry Bush
Euonymus atropurpureus	Burning Bush (Wahoo)
Euonymus obovatus	Running Euonymus (Trailing Wahoo)
Fagus americanus	American Beech
Fraxinus americana	American Ash (White Ash)
Gaultheria procumbens	Wintergreen (Checkerberry) (Teaberry)
Gaylussacia baccata (and other sp.)	Huckleberry
Gleditsia triacanthos	Honey Locust (Black Locust)
Gymnocladus dioica	Kentucky Coffee Tree
Hamamelis vernalis	Vernal Witch-hazel
Hamamelis virginiana	Witch-hazel
Hichoria ovata (Carya ovata)	Shell-bark Hickory
Hydrangia arborescens	Smooth Hydrangea (Seven Barks) (Hills of Snow)
Hypericum prolificum	Shrubby St. Johnswort
Ilex glabra	Inkberry
Ilex lævigata	Smooth Winterberry
Ilex monticola	Mountain Winterberry
Ilex opaca	American Holly
Ilex serrata	Finetooth Holly
Ilex verticillata	Common Winterberry (Black Alder)
Itea virginica	Sweetspire
Juglans cinera	Butternut
Juglans nigra	Black Walnut
Juniperus communis depressa	Prostrate Juniper
Kalmia angustifolia	Sheep Laurel (Lambkill)
Kalmia latifolia	Mountain Laurel
Kalmia polifolia (K. glauca)	Swamp Laurel (Bog Kalmia)
Rhododendron maximum	Rosebay Rhododendron
Rhus canadensis (R. aromatica)	Fragrant Sumach
Rhus Cotinus	Smoke Bush
Ribes aurea	Flowering Currant
Rubus deliciosus	Boulder Raspberry
Rubus odoratus	Flowering Raspberry
Salix discolor (and other sp. unknown)	Pussy Willow
Sassafras variifolium (S. officinalis)	Sassafras
Symphoricarpus racemosus	Snowberry
Symphoricarpus vulgaris var. *Chenaulti*	Coralberry (Indian Currant)
Taxus canadensis	American Yew
Thuya occidentalis	American Arbor Vitae (Yellow Cedar)
Tsuga canadensis	Hemlock
Tsuga caroliniana	Carolina Hemlock
Ulmus americana	American Elm
Ulmus fulva	Slippery Elm
Vaccinium corymbosum	Highbush Blueberry
Vaccinium pallidum	Dryland Blueberry
Vaccinium pennsylvanicum	Lowbush Blueberry
Vaccinium macrocarpum	Cranberry (American)
Vaccinium Vitis-idæa	Cowberry
Vaccinium Vitis-idæa minor	Mountain Cranberry
Viburnum alnifolium	Hobblebush
Viburnum americanum	American Cranberry Bush
Viburnum cassinoides	Withe-Rod
Viburnum Lentago	Nannyberry (Sheepberry)
Viburnum prunifolium	Black Haw
Zanthorrhiza apiifolia	Shrub Yellowroot
Zanthoxylum americanum	Prickly Ash

Trees, shrubs and vines which may be classed as "Herbs" but are not native to the U. S., at Winter Valley, Milton, Massachusetts.

Botanical name	Common name
Acer platanoides	Norway Maple
Acer pseudo-platanus	Sycamore Maple
Æsculus Hippocastanum	Common Horsechestnut
Aristolochia sipho	Dutchman's Pipe
Berberis vulgaris (and other sp.)	Common Barberry
Calluna vulgaris (and vars.)	Heather

Castanea vesca	Chestnut (A Chinese sp. said to be blight-proof)
Corylus Avellana (and hort. vars.)	Filbert
Cratægus Oxyacantha (and other sp.)	English Hawthorn
Daphne Mezereum	Mezereon (February Daphne)
Erica carnea (and vars.) (and other sp.)	Heath
Euonymus europæa	Spindle Tree (European Burning Bush)
Fagus sylvatica	European Beech
Fraxinus Ornus	Manna Ash
Hamamelis mollis	Chinese Witch-hazel
Hedera helix	English Ivy
Hedera helix baltica	Baltic Ivy
Hedera helix conglomerata	Bunchleaf English Ivy
Hedera helix gracilis	Baby English Ivy
Laburnum alpinum	Scotch Laburnum
Laburnum vulgare	Common Laburnum (Golden Chain)
Larix europæa	European Larch
Larix leptolepis	Japanese Larch
Lonicera fragrantissima (and other sp.)	Fragrant Honeysuckle
Malus sylvestris (Hort. vars.)	Apple
Morus tatarica alba	Russian Mulberry
Philadelphus coronarius	Mock Orange
Picea excelsa (and other sp.)	Norway Spruce
Pinus Cembra	Swiss Stone Pine
Pinus densiflora	Japanese Red Pine
Pinus montana	Swiss Mountain Pine
Pinus montana mughus	Mugho Pine
Pinus nigra	Austrian Pine (Indispensable, a joy always)
Pinus nigra var. calabrica (in "Hortus") (*Poiretiana*—in "Stand. Plt. Nms.")	Corsican Pine (A poor, unhappy thing here)
Pinus sylvestris	Scotch Pine
Prunus glandulosa (Hort. vars.)	Flowering Almond
Prunus tomentosa (and other sp.)	Nanking Cherry
Pyracantha coccinea (and var.)	Firethorn
Quercus robur	English Oak
Rhamnus Frangula	Glossy Buckthorn (Elder Buckthorn)
Rosa rubiginosa	Sweet Brier (Eglantine)
Salix alba	White Willow
Salix caprea	Goat Willow (European Pussy Willow)
Salix pentandra	Laurel Willow
Sorbus Aucuparia	European Mountain Ash (Rowan Tree)
Styrax japonica	Japanese Styrax
Syringa vulgaris (and Hort. vars.)	Common Lilac
Taxus baccata repandens	Spreading English Yew
Ulmus campestris	English Elm
Viburnum Carlesii	Fragrant Viburnum
Viburnum fragrans	Fragrant Viburnum
Viburnum Opulus	European Cranberry Bush
Viburnum Opulus sterile	Common Snowball
Vitex macrophylla	Chaste Tree
Vitex negundo	Negundo Chaste Tree
Vitex negundo incisa	Cutleaf Chaste Tree

GREENHOUSE HERBS

Osmanthus fragrans (*Olea fragrans*)	Fragrant Olive
Osmanthus Aquifolium	Holly Osmanthus
Thea sinensis (*Camellia thea*)	Tea
Viburnum tinus	Laurustinus

Herbs at Little Harbor Farm, Woods Hole, Massachusetts. (As of September, 1934.)

Agrimony	*Agrimonia Eupatoria*
Angelica	*Angelica archangelica*
Catsfoot (Pussytoes)	*Antennaria dioica*
Cedar, Red	*Juniperus virginiana*

Common name	Botanical name
Avens	*Geum urbanum*
Balm	*Melissa officinalis*
Balm of Gilead (Mosquito Plant)	*Cedronella cana* (or *C. triphylla*)
Basil, Bush	*Ocimum minimum*
Basil, Purple Bush	*O. comosum*
Basil, Sweet	*O. Basilicum*
Bedstraw, Hedge	*Galium Mollugo*
Bedstraw, Ladies'	*G. verum*
Bergamot	
(Bee Balm) (Oswego Tea)	*Monarda didyma*
(Salmon Pink)	*M. didyma* var. *salmonea*
(Pink)	*M. kalmiana*
(White, Tall)	*M. superba*
(Virulent magenta)	*M. media*
Betony, Wood (Woundwort)	*Stachys betonica* (*Betonica officinalis*)
Betony, Big	*Betonica grandiflora*
Betony, Woolly	*Stachys lanata*
Bittersweet	*Solanum Dulcamara*
Blessed Thistle	*Cnicus benedictus*
Borage	*Borago officinalis*
Box (Truedwarf)	*Buxus sempervirens* var. *suffruticosa*
Broom	*Cytisus scoparius* (*Genista scoparia*)
Bryony	*Bryonia dioica*
Burnet (Lesser or Salad)	*Sanguisorba minor* (*Poterium sanguisorba*)
Calamint, Alpine	*Satureia alpina* (*Calamintha alpina*)
Calamint, Lesser	*S. Nepeta* (*C. Nepeta*)
Camomile, English	*Anthemis nobilis*
Camomile, German	*Matricaria chamomilla*
Camphor Plant	*Chrysanthemum balsamita* (*Balsamita vulgaris*)
Catnip (Cat Mint)	*Nepeta cataria*
	*N. lanceolata** (White flowers)
	*N. nepetella** (Mauve fls.—tall, interesting plant resembling a refined and slender Clary)
	N. mussinii
	N. nervosa
	N. "Souvenir d'Andre Chaudron"
Chaste Tree	*Vitex Agnus-castus*
Chervil	*Anthriscus cerefolium*
Chives	*Allium schœnoprasum*
Cinquefoil	*Potentilla reptans*
Clary	*Salvia sclarea*
Clove Pink	*Dianthus caryophyllus* ("Grenadine Red")
Clover, Yellow Sweet (Melilot)	*Melilotus officinalis*
Cornflower	*Centaurea cyanus*
Corn Salad	*Valerianella olitoria*
Costmary (Alecost) (Bible Leaf)	*Chrysanthemum balsamita tanecetoides*
Cupidone (Cupid's Dart)	*Catananche cœrulea*
Currant, Black	*Ribes nigrum*
Currant, Red	*R. rubrum*
Currant, White	*R. rubrum* var. *album*
Dalmatian Powder Plant	*Chrysanthemum cinerarifolium* (*Pyrethrum cinerarifolium*)
Dill	*Anethum graveolens*
Dyers' Greenwood (Woadwaxen)	*Genista tinctoria*
Echinacea (Cone Flower)	*Echinacea angustifolia*
Echinacea (Purple Cone Flower)	*E. purpurea*
Edelweiss	*Leontopodium alpinum*
Elder, Dwarf	*Sambucus Ebulus*
Elecampane	*Inula Helenium*
Eryngo (Sea Holly)	*Eryngium amethystinum*
Flax, Perennial	*Linum perenne*
Foxglove	*Digitalis purpurea*
Garlic	*Allium sativum*
Geranium, Lemon	*Pelargonium crispum*
Geranium, Lemon	*P. capitatum*
Geranium, Nutmeg (Apple Geranium)	*P. odoratissimum*
Geranium, Oak-leaved	*P. quercifolium*
Geranium, Peppermint	*P. tomentosum*
Geranium, Rose (and 2 vars.)	*P. graveolens*
Germander	*Teucrium Chamædrys*
Goats' Rue	*Galega officinalis*
Gooseberry, Red	*Ribes grossularia*

Gooseberry, White.......... *R. grossularia*

Holly, English.............. *Ilex aquefolium*
Hops........................ *Humulus lupulus*
Horehound (or Hoarhound).. *Marrubium vulgare*
Hyssop...................... *Hyssopus officinalis*

Indigo...................... *Indigofera tinctoria* (Tropical plant—treat as an annual)

Juniper, Savin.............. *Juniperus sabina*

Lavender, Spike............ *Lavandula spica*
Lavender, Sweet............ *L. vera*
Lavender, ——?............ *L. pinnata* (Tender, not hardy through winter)
Lavender Cotton............ *Santolina Chamæcyparissus incana*
Lemon Verbena.............. *Lippia citriodora*
Lily, Madonna............... *Lilium candidum*
Lime Tree (Linden).......... *Tilia vulgaris* (*Tilia europæa*)
Locust, Common............. *Robinia pseudoacacia*
Lovage...................... *Levisticum officinale*

Mallow, Common or Blue.... *Malva officinalis* (*M. sylvestris*)
Marigold, Pot............... *Calendula officinalis*
Marjoram, Pot.............. *Origanum onites*
Marjoram, Sweet............ *O. Majorana*
Marjoram, Wild............. *O. vulgaris*
Marshmallow................ *Althæa officinalis*
Meadowsweet............... *Spirea Ulmaria* (*Filipendula Ulmaria*)

Melilot, Blue................ *Melilotus cærulea* (*Trigonnella cærulea*)

Mints
Curly Mint.............. *Mentha crispa*
———?................ *M. perfoliata saxonis*
Peppermint.............. *M. piperita*
Pennyroyal.............. *M. Pulegium*
Applemint.............. *M. rotundifolia* (Leaves dark green and very much rounder than those of any of the other mints)
Applemint (Variegated).... *M. rotundifolia*, var. *variegata* (Leaves variegated grayish-green and white, much crinkled and rather woolly.) Both of these smell very decidedly of apples and it is only when they are crushed that the mint fragrance becomes uppermost. These tally with the descriptions of each, and their names, both botanical and English, as given in "Hortus."
Spearmint (Lamb's Mint) .. *M. spicata* (*M. viridis*)

Thymes
*Thymus azoricus**
T. cimicinus
T. erectus
Caraway-scented Thyme... *T. herba-barona* (Charming, delicious)
T. hyemalis
T. marshalliensis
*T. micans**
T. nitidus (Charming, very useful in sunny wild garden)
T. nummularius (Fine, robust, spreading plant)
Thyme, Wild (Mother of Thyme)............ *T. Serpyllum**
Thyme, White............ *T. Serpyllum albus*
T. Serpyllum carnosus
Thyme, Lemon-scented.... *T. Serpyllum citriodorus*
T. Serpyllum coccineus
Thyme, Woolly........... *T. Serpyllum languinosis*
T. Serpyllum minus (A treasure—tiny, flat, very dark green)
T. Serpyllum odoratissimus
T. Serpyllum pulchellus
T. Serpyllum roseus
T. Serpyllum villosus
T. Serpyllum zygis
Thyme, Broad-leafed Eng.. *T. vulgaris*
Thyme, French........... *T. vulgaris* (No special botanical name given for this—a delightful heatherlike plant)
Thyme, Silver............ *T. vulgaris argenteus*
T. vulgaris fragrantissimus

There is so much confusion about Thymes and so few definite descriptions given to distinguish one from another that these last four seem to require some notes. With us they are as follows:

T. vulgaris "Broad-Leafed English" Common Thyme: Leaves rather dark green. Fls. inconspicuous white. Ht. 6". Dies out in patches after flowering and must be severely pruned.

T. vulgaris "French variety." Leaves dark gray-green. Fls. same as above. Ht. 8-10". Robust, bushy plant almost exactly like our T. fragrantissimus and in the distance much resembling heather.

T. vulgaris argenteus: Again a lusty shrublet like the above in height. Variegated light gray-green and white. Not

Lemon Mint (This English name is according to Mrs. Bardswell who gives no botanical name) .... *M. sp.* (Leaves variegated green and yellow and more smooth and shining than those of any of our other mints)
Mugwort .... *Artemisia vulgaris*

Nasturtium .... *Tropæolium majus*

Parsley .... *Petroselinum hortense*
Passion Flower (Purple granadilla) .... *Passiflora edulis*
Pellitory-of-the-Wall .... *Parietaria officinalis*
Plum, Beach .... *Prunus maritima*
Plume Poppy .... *Bocconia cordata*
Poppy, Opium .... *Papaver somniferum*
Psylla .... *Plantago Psyllium*

Rampion .... *Campanula Rapunculus* (or *C. rapunculoides*) (have got seeds of both twice over and the plants from each seem to be both exactly alike)
Rosemary .... *Rosmarinus officinalis*
R. officinalis "Heavenly Blue"
Rue .... *Ruta graveolens*

Safflower (False, or Bastard, Saffron) .... *Carthamus tinctorius*
Sage .... *Salvia officinalis*
Savory, Summer .... *Satureia hortensis*
Savory, Winter .... *S. montana*
Scurvy Grass .... *Cochlearia officinalis*
Soapwort (Bouncing Bet, etc.) *Saponaria officinalis*
Sorrel .... *Rumex acetosa* (or *R. patientia*—have both sp.)
Southernwood (Old Man) (Lads' Love) .... *Artemisia Abrotanum*
Stonecrop .... *Sedum acre* (and other sp.)
Sweet Cicely .... *Myrrhis odorata*

Tansy, Huron .... *Tanecetum huronensis*
Tarragon .... *Artemisia dracunculus*
Toadflax——? .... *Linaria ericoides Hendersoni*
——? .... *Linaria ericoides* "Canon J. Want"
(A delightful plant for border, rock garden or anywhere. Foliage grayish and heathlike, Fls. salmon pink).
——? .... *L. pallida*
——? .... *L. pallida alba*
Ivyleaved Toadflax (Wall Pennywort, Kenilworth Ivy) .... *L. Cymbalaria*
Tobacco, Aztec .... *Nicotiana rustica*

argenteus. Have seen one English catalogue which lists both kinds of "Silver Thyme" but otherwise our kind seems never to be mentioned.
T. vulgaris fragrantissimus: Almost exactly like our French Thyme. A little grayer and a little more fragrant.

Veravin .... *Verbena officinalis*

Willow Herb .... *Epilobium angustifolium*
Woad .... *Isatis tinctoria*
Wormwood, Common .... *Artemisia absinthinum*
Wormwood, Beach .... *A. stelleriana*
Wormwood, Roman .... *A. pontica*
Wormwood, Saw-toothed .... *A. serrata*
Wormwood, "Silver King" .... *A. albula*

WEEDS: (Dug up on the place and transplanted to top of wall above herb garden.)
Clover, Alsike .... *Trifolium hybridum*
Clover, Low Hop .... *T. procumbens*
Clover, Rabbit's Foot .... *T. arvense*
Clover, Red .... *T. pratense*
Clover, White .... *T. repens*
Cranesbill (Small-flowered) .. *Geranium pusilium*

Feverfew .... *Chrysanthemum Parthenium*

Milkwort, Racemed (Bitter Polygala) .... *Polygala polygama*

Pimpernel, Scarlet (Poor Man's Weather Glass) .... *Anagallis arvensis*

Cedar, White .... *Chamæcyparis thyoides* (Little seedlings growing wild in moss in wall chinks, *not planted* there)
Liverwort (True) .... *Marchantia polymorpha* (*Not* put there on purpose—it just appeared)

* Asterisks after certain Nepetas and Thymes are to explain that the plants we have under these names actually do differ one from the other although, according to "Hortus," these different names apply to one and the same sp. Perhaps we were sent wrongly marked seeds or perhaps our labels got mixed.

This herb garden contains also 55 genera and 10 sp. of plants and 15 sp. of trees or shrubs which are not listed here because they are both happier at and more appropriate to our Milton place and so they are on the latter's list instead.

PLANTS
for the
HERB GARDEN

You will want a copy of our booklet "Plants for the Herb Garden," which includes over sixty varieties of specialized plant materials with which to create a green garden of use and delight.

Many of these plants are practical for their flavoring, seasoning or medicinal properties: Some, like Rosemary and Lad's Love, carry a cargo of fragrance and sentiment.

Others, like Germander, Lavender Cotton and Hyssop, add their neat foliage to the group, particularly in trim "Knot-gardens." A copy of this booklet will be mailed to you on request.

Let us remind you, too, that we grow Evergreens, Trees, Flowering Shrubs, Vines, Roses, and Perennials in a wide variety, enabling us to meet all your garden requirements. These are illustrated and accurately described in our 96-page Handbook for 1935 — also mailed on request.

BAY STATE NURSERIES

Incorporated

NORTH ABINGTON MASS. — also — FRAMINGHAM MASS.

Telephone Rockland 26 — Framingham 6191

The HERBARIST

A Publication of the
Herb Society of
America

For Use and for Delight

BOSTON, MASSACHUSETTS

1936

CONTENTS

THE PLATES on the following two pages are reproductions of plates 7 and 5 of the Badianus manuscript.

PLATE 7. *Huacalxochitl — Xanthosoma Roseum Schott (Standley Bot. Ser.* V, X, *p.* 124)

Aquatic plant of the Arum family no longer found at the altitude of Mexico City although fairly common in the lowlands.

PLATE 5.

The plant Tlapalcacauatl, shown in the middle of the top row, is translated as "colored cocoa." This is one of two varieties shown in the manuscript. From this the Aztecs derived their chocolate.

Both plates are shown by courtesy of the Bulletin of the Institute of Medicine, *Vol.* III, *No.* 6, *June* 1935.

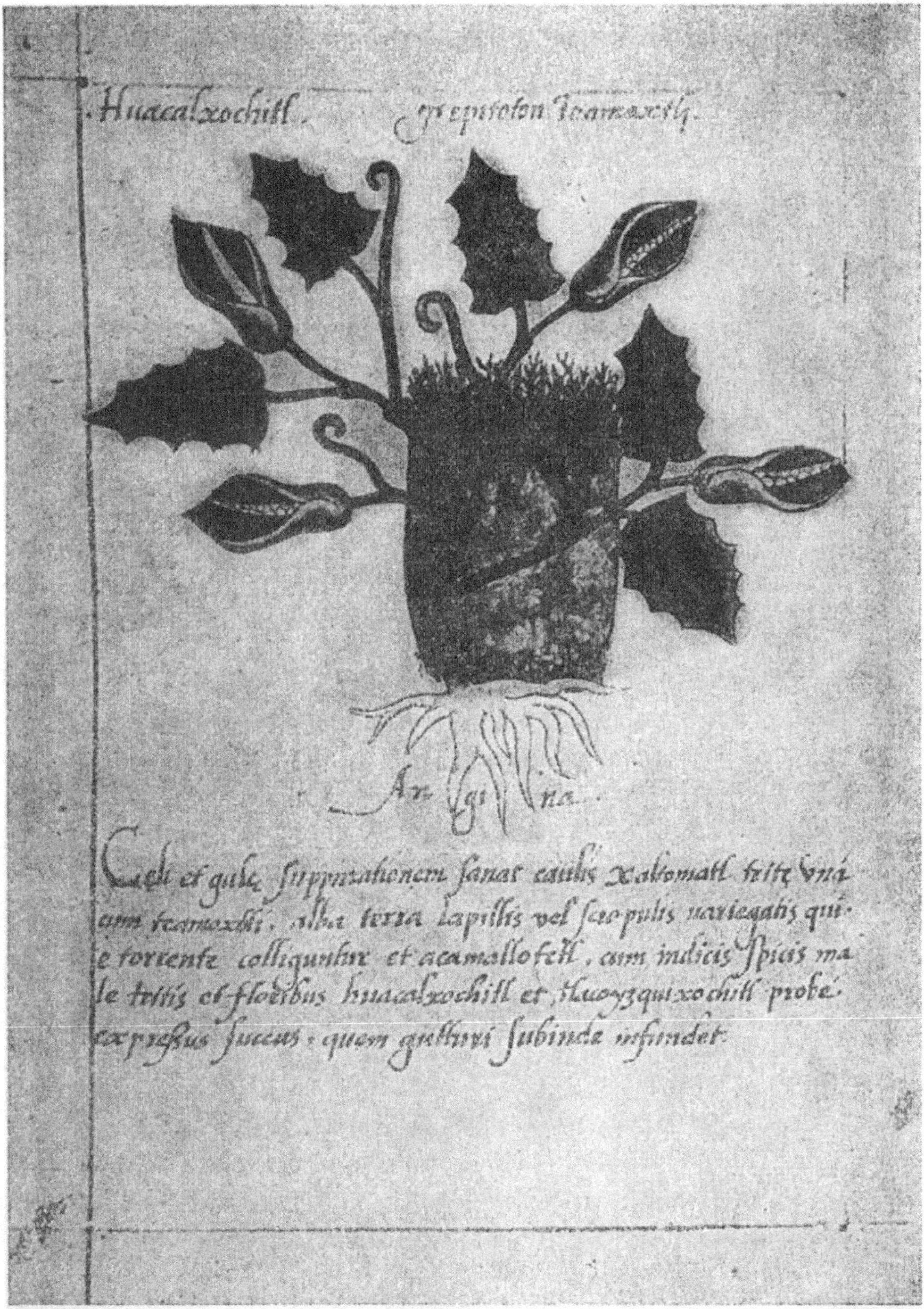
Huacalxochitl.
An gi na
Caeli et gulae suppurationem sanat caulis xaltomatl tritus una cum tlamoxcalli, alba terra lapillis vel scrupulis variegatis qui e torrente colliguntur et acamallotetl, cum indicis spicis male tritis et floribus huacalxochitl et tlaoyzquixochitl probe expressus succus quem gutturi subinde infundet.

Plate 7

Temahuiztiliquauitl. Tlapalcacauatl. Texcalamacoztli.
Couaxocotl. Yztacquauitl. Tevezquauitl. huitzquauitl.

Plate 5

AN OLD HERBAL "NEWE FOUNDE"

ALL lovers of old books and all interested in Herbal lore will receive with delight the "Joyfull Newes" of the recovery from comparative oblivion of an authentic Aztec Herbal of the year 1552.

Previously known to only a few scholars, and obscured under the cryptic title of "*Codex Barberini, Latin 241*," this manuscript, written in Aztec and in Latin by two native pupils of the Franciscan friars of Mexico City, has at last been brought back, in facsimile at least, to the same continent from which it originated so many years past.

About five years ago, through the generosity of Ambassador Charles G. Dawes, Dr. Charles U. Clark was sent to Europe to search for early Latin-American texts. In the Vatican Library he unearthed a unique document known as the Badianus manuscript and through the courtesy of the Library was enabled to bring back photographs of the original pages and a set of color sketches, accurately copied from the original illustrations.

From these photographs Dr. Emily Walcott Emmart, of The Johns Hopkins University, has made an English translation, a labor of four years' intensive research, the value of which can scarcely be overestimated. With a knowledge of medicine, language, and botany at her command, she has based her translation on the etymological sources of the Aztec words and has cross-referenced every Aztec plant name with 16th century Aztec-Latin texts. What is most illuminating of all, she has identified them for us according to modern botanical description.

Dr. Emmart has written a thrilling *résumé* of the manuscript's contents and the history of its compilation,* which is a foretaste of joys to come, when the Herbal and translation shall be published.

She writes in description: "The manuscript is a complete Herbal, consisting of 63 folios, approximately 6 by 8¼ inches in size, clearly

*See *Smithsonian Miscellaneous Collections,* Vol. 94, No. 2, Publication 3329, May 18, 1935, and *Bulletin of the Institute of the History of Medicine,* Vol. III, No. 6, June, 1935.

written in Latin and Aztec. . . . The Text is exquisitely illustrated with pictures of 204 herbs and trees and these illustrations still today retain their brilliancy of color. . . . The Herbal is the work of two Aztecs who were educated at the College of Santa Cruz, Mexico City. It was first written in Aztec and then, with the exception of the names of plants, stones and animals, translated into Latin within the same year.

" The exact title reads as follows:

A book of Indian Medical Herbs
composed by a certain Indian physician of the
College of Santa Cruz, who is not theoretically learned,
but is taught only by experience.
In the year of our LORD Saviour
1552.

" There seems little doubt that the principal author of the manuscript is one Martin de la Cruz, whose name appears in the first line of the dedication; the second author is Jaunnes Badianus, the translator, whose signature appears in the postscript at the end of the last chapter. . . . The Badianus manuscript holds the unique position of being the earliest written Aztec Herbal, as well as being the only one written by the Aztecs themselves. . . . It marks the beginning of Herbal literature on this side of the Atlantic."

Herb enthusiasts and botanists will await with ill-restrained eagerness the publication of this triumph of research and will welcome the day when their minds " may be singularly enriched with the knowledge of these visible things."

The publication of the complete facsimile of the Badianus manuscript, with 92 colored plates, involves a considerable sum of money, the whole of which the Smithsonian Institution is unable to supply unaided. Dr. C. G. Abbot, Secretary of the Institution, authorizes the statement that if $6,000 can be raised from outside sources, the Smithsonian Institution would see its way to publish the Badianus manuscript in facsimile together with Dr. Emmart's annotated translation. If there are those who would be interested to contribute toward the publication of this valuable manuscript will they communicate with

Dr. C. G. Abbot, Secretary, Smithsonian Institution
Washington, D. C.

1

Libellus de medicinalibus Indorum her-
bis, quem quidam Indus Collegii sanctę
Crucis medicus composuit, nullis
rationibus doctus, sed solis ex-
perimentis edoctus. Anno
domini seruatoris
1552.

Clarissimo domino Francisco de Mendoza
illustrissimi superioris huius Indię proregis
domini Antonij de Mendoza filio optimo Mar-
tinus de la Cruz, indignus seruus salutē
precatur plurimam & prosperitatem.

+

ex libris didaci Cortavila

Cum tu omnium uirtutum decora atq ornamenta & bonorū
munera, quę a quouis mortali desiderantur reluceant domine
magnificentissime. nescio profecto. quid in te potissimum lau-
dem. Equidem non uideo. quibus laudibus efferam insignem
uestrum amorem: quibus uerbis gratias agam uestro beneficio
quā maximo. Pater enim tuus Vir christianissimus simul et pijs-
simus, uerbis consequi nequeam quā maxime omnium mihi
benefecerit. quicquid enim sum, quicquid possideo & quicquid
nominis habeo, illi debeo. Nihil par, nihil ne dignum inue-
nire

FACSIMILE OF THE FIRST PAGE OF THE BADIANUS MANUSCRIPT,
"CODEX BARBERINI, LATIN 241" (VATICAN LIBRARY)

meris imposuit. Vale in Christo Seruatore. Tlatilulci in Sanctę Crucis Collegio Diuę Marię Magdalenę sacratis ferijs. Anno ab orbe restituto Milleßimo quingenteßimo quinquageßimo secundo.

Finis libelli herbarij, quem latinitate donauit Joannes Badianus natione Indus patria Xuchimilcanus eiusdem collegij prelector.

Gloria semper ei sit cuius munere uersi,
Quem cernis librum lector amice bone.

FACSIMILE OF THE LAST PAGE OF THE BADIANUS MANUSCRIPT

THE TARRAGONS, CULTIVATED AND WILD

EDGAR ANDERSON

TWO distinct tarragons are now on the market in this country. One is sterile and does not produce seed; it must be propagated vegetatively. The other can be grown from seed but it is vastly inferior as a culinary herb. In the garden these two tarragons look very much alike; even to the trained botanical eye there is little outward difference between them, but to a real herbarist they belong in quite separate categories, for the sterile one is among the best salad herbs; the fertile one has a rank, disagreeable taste. In horticultural and botanical treatises, as well as in catalogues and nurseries, the two have been much confused.

Nature herself has made the problem difficult. All tarragons belong to a species allied to the wormwoods, *Artemesia Dracunculus,* a notoriously variable species of a notoriously difficult genus of plants. By modern botanists *Artemesia Dracunculus* is understood to include a vast group of plants native all the way from our own great plains and Rocky Mountains across Siberia to the eastern edge of Europe. With such a distribution it is scarcely surprising to learn that the species is a variable one. Some varieties are hairy, others are smooth. Some have a rank odor, some have no odor at all. It is nearly as complex a species as mankind itself which includes Blondes and Brunettes, Negroes, Eskimos, Malays, and Mongols in a complex grouping of races, subraces and tribes that defies exact and orderly classification. Among all the varying forms, Asiatic and American, which belong to the species *Artemesia Dracunculus,* there is apparently but one which has superlative qualities as a domesticated herb. This is the tarragon which may occasionally flower but never sets seed, the tarragon which is often known as " French tarragon " or " German tarragon " but which really was anciently domesticated somewhere in central Asia and did not reach Europe until towards the end of the Middle Ages.

Of the wild tarragons several may well be in cultivation at one place or another, though only one strain has been seen by the writer. This is a rank-flowered variety sometimes known as Russian tarragon or Siberian tarragon. In addition to its flavor and its ability to set seeds it has a few trifling differences from the cultivated variety. It is first of all more variable from plant to plant. On the average, the margins of the leaves are a little heavier and a little wider and there are a few scattered hairs. These are so few that one can hardly speak of the leaves as hairy. Though the soft white-hairs are easily seen with a hand lens there are seldom more than one or two to a leaf. The flower heads are furthermore a little larger in these wild tarragons and the stalks below the heads tend to be somewhat longer.

When one searches for the precise and correct botanical names for these two varieties which are now in our gardens he finds man-made confusion added to the original difficulties presented by nature herself. So few horticulturists and botanists have been inclined to go to the root of the matter that the confusion in the books is worse than that in the gardens. As early as 1749 Gmelin, in quaint but descriptive Latin, had detailed his troubles in classifying the variable Siberian tarragons. The Russian botanist, Ledebour, put the Siberian tarragons in a separate species and called them *Artemesia Redowski.* Willdenow likewise thought them worthy of a separate specific name and named them *Artemesia inodora.*

Few botanists today would agree with Ledebour and Willdenow in classifying the wild and cultivated tarragons in separate species. The majority would follow Besser, the Russian botanist who a century ago published a series of papers of great technical brilliance dealing with this difficult genus. The cultivated tarragons he classified as *Artemesia Dracunculus* L. var. *sativa* Besser. The more variable wild sorts he considered to be another variety of the same species, *Artemesia Dracunculus* L. var. *inodora* (Willd.) Besser. Unfortunately at the time he did his work there were not, as there are today, sets of international rules by which botanists decide upon the exact name to be used in such cases. This means that when some modern student takes up the problem where Besser left off, and gives us a comprehensive treatment of the

Asiatic Artemesias, the latter of these names will probably have to be changed again for purely bibliographical reasons. Until that time we had best use Besser's names, *Artemesia Dracunculus* var. *sativa* and *Artemesia Dracunculus* var. *inodora,* adding to them the identifying abbreviations for the authors [as was done above] when we wish to achieve the greatest possible precision.

The cultivated tarragon is only one of several instances in which a sterile but horticulturally superior variety of a garden herb has been spread around the world by vegetative propagation. The horseradish of our gardens may flower, but it never sets seed; if one wishes to see a fertile plant he must journey to central Asia, the ancient home of that herb. More than one of the finest watercresses is quite sterile and cannot be grown from seeds. Which means, of course, that when we take a plant of the true tarragon into our gardens and grow it, and at length divide the root, we are the last link in a long chain of such people. Just so did the man from whom we had the herb, and the man from whom *he* had it, and the man from whom *he* had it, and the man . . . It was indeed by a series of just such steps that the cultivated tarragon spread out of its original Asiatic home and across central Europe and finally around the world.

"Ringing Bees"
Matthiolus, Petrus Andreas
Commentarii, in libros sex Pedacii Dioscoridis
Venetiis, In Officina Valgrisiana, 1560.

OLD BEE LITERATURE

ELLEN GREENSLET

BEGINNING with Aristotle, the earliest author whose treatise on bees has come down to us, it is amazing to see how much the ancient Greeks and Romans had been able to learn without the aid of the microscope.

Aristotle studied his bees, probably, in the Greek Isles where he lived, while writing his great treatise *On Animals*. Perhaps again, later, in that Athens garden — surely one of the pleasantest gardens in all literature — where he taught his students and spent long hours with his friend and pupil Theophrastus, whose work on plants earned for him the title of " Father of Botany."

Almost every other Greek author of note had allusions to the bees, and we find several poems in the Greek Anthology. One of these, entitled in Mr. Mackail's translation, *Loca pastorum deserta*, is interesting as showing the very early appearance of that common mediæval belief, persisting even up to the last century, that the bees must be told if death comes to their owner or one of his family, otherwise they will forsake their hives and grieve to death.

> "Naiads and chill cattle-pastures, tell to the bees when they come on their springtide way, that old Leucippus perished on a winter's night, setting snares for scampering hares, and no longer is the tending of the hives dear to him; but the pastoral dells mourn sore for him who dwelt with the mountain peak for neighbor."

There is a modern version of this same belief in a poem written nearly two thousand years later:

> "Just the same as a month before —
> The house and the trees,
> The barn's brown gable, the vine by the door —
> Nothing changed but the hive of bees.
>
> Before them, under the garden wall,
> Forward and back,
> Went drearily singing the chore-girl small,
> Draping each hive with a shred of black.
>
> And the song she was singing ever since
> In my ear sounds on:—
> 'Stay at home, pretty bees, fly not hence!
> Mistress Mary is dead and gone.'"

We get another glimpse of a bee garden in the Anthology — a tragic garden — of which we know neither its owner's name nor the place — only that the little baby, Hermonax, creeping toward the hives in quest of honey, was attacked by a savage pack of bees and stung to death.

Another poem tells of "the old bee-keeper Cliton," who, desirous of making an offering to some rustic divinity, preferred to give the fruit of his own toil rather than the accustomed victim from the herd, and so "cut out and offered a honeycomb, the sweet harvest of his swarm — the gift of his unshepherded, far-flying flock."

Several hundred years later, the Roman poet Tibullus, in his beautiful poem "Against War," after a few lines in which he says that war comes because men no longer have to use their weapons against wild beasts, and so "turn them to mischief" against each other, concludes the passage by recalling the simple, busy life of his grandfather's time. He tells of the "wooden god, standing poorly garbed in a narrow shrine — whose favor was won when a man offered a bunch of grapes as first fruits, or laid the spiky garland on his hoary hair, and one who had gained his prayer would, with his own hands, bring the honey cake, his little daughter following with the pure honeycomb in hers." This picture of a little girl from a Roman farm and the old Greek bee-keeper, each making so simple and humble an offering to their rustic gods, an offering taken from their own dooryards, is not least of the many pleasant memories one brings away from excursions through the ancient literature of the bee.

As a rule we must, in later literature, look for bee-gardens in books on husbandry. The first Roman description of one is found in a book written in the first century B. C. by Varro, a very learned Roman, and neighbor to Virgil, at his villa outside of Naples. When an old man of eighty, he gathered together and condensed all the learning of past ages on agriculture in his *Rerum Rusticarum Libri.* This he dedicated to his wife, Fundania, in the hope that after his death its teachings would enable her to continue to manage their country estates. Many Greek and Latin authors that have since been lost were quoted, their teachings summarized and arranged, and the whole presented in dialogue form, to make it more popular in its appeal, and better adapted to the use of the thousands of soldiers just returned from the wars and settled on small holdings all over Italy.

In this book, Varro, in telling of the brothers Veiani, two retired soldiers who had served under him in Spain, gives the first picture of a bee-garden in Roman literature. This was a little plot with "bee hives set all about the house (a small house with a curtilage of little more than half an acre about it) with part of the land planted in a garden, and the rest filled up with thyme and clover, and that bee plant called '*melittæna*'; from this garden the owners were wont to derive an average income of not less than ten thousand sesterces a year from honey"; (about five hundred dollars). But Varro sagely adds they did this by "being willing to wait until they could sell at their own time and price."

The great interest in the Veiani's garden, however, lies not in its description, but in the fact that it was quoted — along with the one later described in Virgil in the Georgics — for well over fifteen hundred years in practically every treatise that was written on bees. Barnaby Googe, in translating into English in 1614, the work of a Dutch author, Conrad Heresbach, introduced them to England.

At about the time Varro wrote his *Rerum Rusticarum*, great pressure was brought to bear on Virgil — already famous for the reception accorded his Eclogues and a great favorite of the Emperor Octavius — to write a poem on husbandry, the government hoping that with such illustrious sponsors as Varro and Virgil, their back-to-the-farm policy, which they were trying to enforce on rich and poor alike, would be carried through to success.

Throughout his childhood, Virgil had lived on a farm, and was thus familiar with farming operations and bee-keeping. The practical part of his poem, however, was based almost entirely on Varro's teachings. The Georgics, a poem of about two thousand lines, which Virgil composed with the greatest care, is one of the very great poems of all literature, and is held by many critics to be the most perfect poem in the world. The fourth and last book treats of bees, the first being devoted to tillage, the second to planting, and the third to the rearing of cattle. Virgil wrote this poem at his winter home, his beloved " Sweet Parthenope," overlooking the Bay of Naples, perhaps in its garden, filled with the buzzing of beautiful golden bees — that Italian breed which has been kept so pure through the centuries that it is, even today, universally proclaimed the aristocrat of the hive.

Virgil begins the fourth Georgic with a brief statement of its theme, and then describes the place where the bees are to be kept, the outstanding description of a bee-garden in all literature, setting a pattern which has been followed for nearly two thousand years:

> " First," he says, " seek a settled home for your bees, whither the winds may find no access. . . . Let clear springs be near, and moss-green pools, and a thread of water stealing through the grass; and let a palm or tall wild olive shade the porch, so that when the new king leads forth the early swarms in the spring they love, . . . a bank nearby may tempt them to retire from the heat, and a tree in their path may hold them in its sheltering leafage. In the midst of the water, whether it stands idle or flow onward, throw willows across, and large stones that they may have frequent bridges on which to settle and spread

> their wings to the summer sun. . . . All about let green cassia bloom, and wild thyme with fragrance far borne, and a wealth of strong-scented savory, and let violet-beds drink of the trickling spring."

Later in the poem, after touching on various subjects such as swarming, the battles of the queens (or kings as they were always called in classic literature), the diseases of the bees were described. Virgil's prescription is the charming predecessor of many similar recipes which the old bee-keepers compounded in an effort to aid their ailing charges.

Toward the end of the poem, we have a glimpse of another bee-garden, that of " the old Corycian," an ex-pirate who had been forced to settle at Tarentum on a few acres of unclaimed land, with a soil

> " not rich enough for bullocks sowing, unfitted for the flock, and unkindly to the vine. Yet, as he planted herbs here and there among the bushes, with white lilies about, and vervain, and slender poppy, he matched in contentment the wealth of kings, and, returning home in the late evening, would load his board with unbought dainties. He was first to pluck roses in spring, and apples in autumn; and when sullen winter was still bursting rocks with the cold, and curbing running water with ice, he was already culling the soft hyacinth's bloom, chiding laggard summer and the loitering zephyrs. So he, too, was first to be enriched with mother-bees and a plenteous swarm, the first to gather honey as it frothed out from the squeezed comb."

Within the next hundred years — that is, during the first century A. D. — two other important works on bees appeared, those of Pliny and Columella. Pliny's treatise appeared in his long and encyclopædic *Naturalis Historia,* a work compounded of fancy and fact which, though full of what Maeterlinck aptly calls " Erreurs Charmantes," is still to this day a mine of interesting information. In speaking of hives, he mentions the use of " lantern horn," evidently some kind of mica, the first (though unsuccessful) expedient to enable the bee-keeper to observe the interior life of the hive. Pliny, for all he wrote so much about it, has never ranked as one of the great writers on husbandry, that part of his book being overshadowed by the mass of other material of which he treats.

Columella, on the contrary (perhaps for the reason that the other writers before him were all known to fame because of some other more noteworthy accomplishment in letters), achieved a reputation as *the* preeminent agricultural writer of Roman times. He was a Spaniard who came to live in Rome in early youth. He was a great admirer of Virgil and, with the exception of Palladius, who followed two hundred years later, was the most quoted writer on his subject until the time of Crescentius in the twelfth century. He treats of the management of bees in great detail, and stresses the advantage of having the bees kept adjacent to the house. "The hives may be made of bark of cork-tree or of giant fennel woven together," or "join willows together, and weave them like weavers work: and, if these cannot be had, they must be made of the wood of a tree made hollow, or sawn into boards."

Columella gives long lists of plants suitable for the bee-yard. These lists, followed by practically every writer after him, differ in different countries, according to the local flora, but thyme, winter savory, rosemary, and marjoram are always present.

Palladius is even today an obscure person of whom little is known. His Latin was dull and heavy in style; but, perhaps because he threw his material into a convenient form for reference, dividing it into calendars of work to be done each month, his book had the vitality to live on through the Middle Ages; and attained the distinction of being used as a sort of horn book by the mediæval monks all over Europe.

Milton had so high a regard for these Latin writers on husbandry — Varro, Columella and Palladius — that when he was tutor to his sister's two sons, he made them read all their works in the original Latin.

There was very little writing on bees during the Dark Ages. Charlemagne, however, in the *Titulaire de Villis* (*Vel Curtis*), of 812 A. D. (the same edict which contains the list of herbs so valuable to us as giving some idea of the plants common to the small enclosures during the early Middle Ages), gives various specific directions as to the amounts of honey and wax that were to be furnished as tithes to his agents by the small farms at definitely stated intervals. These inspectors, who reported everything back to the Emperor in the greatest detail, superintended with the most meticulous care every activity throughout the whole extent of his vast, heterogeneous empire, and small bee-gardens must surely have been dotted all over Europe from Brittany to the Danube. It is known that Charlemagne himself, at his various farms, had sixty

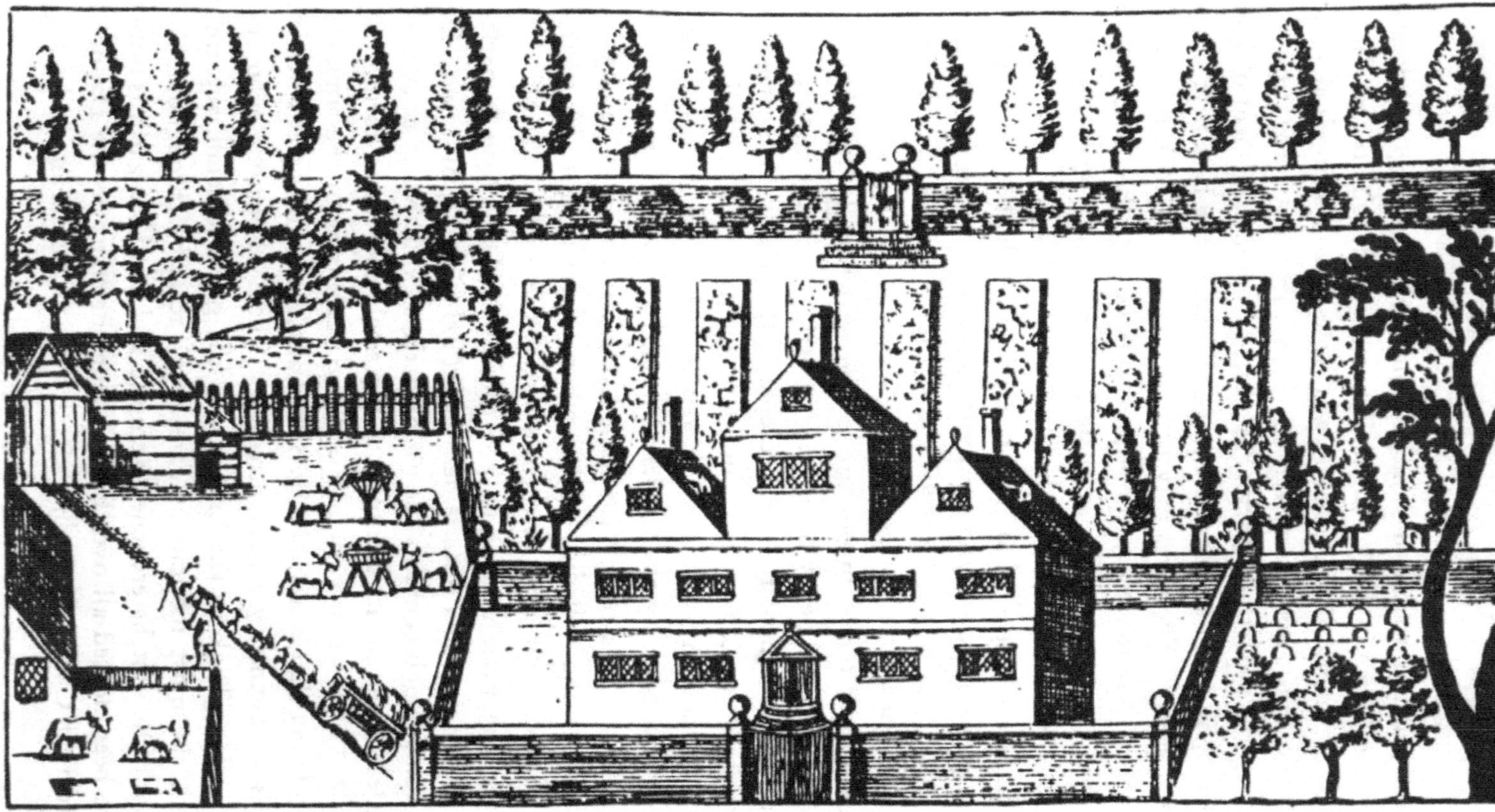

Garden to farmhouse of 1681, with bee-garden adjoining the house

Worlidge, J. Systema Agriculturæ, The Mystery of Husbandry Discovered. (London, 1681) 3rd Edition.

bee hives in which, as in all other agricultural occupations, he took the most lively and personal interest.

A learned Moor, Ibn-al-Awam, who lived at Seville in the twelfth century, wrote a book on agriculture, *Kitab al-felalah,* which fortunately can be studied from a modern French translation. This book, like so many of those published later in England, gave instructions for the "everyday life of a country gentleman," culled largely from Nabathean, *i.e.*, Arabic, Chaldean, Persian, and Egyptian sources. His remarks on bees are interesting as showing the Arab and Spanish usages, and he gives what he affirms is an efficacious cure for stings.

About the middle of the tenth century, the Byzantine *Geoponica*, a compilation of ancient Greek and Latin agricultural treatises, was made for the Emperor at Constantinople. This treats of flowers, pot herbs, and domestic animals, including bees. The *Geoponica* has a high regard for Varro, and quotes from various Greek authors all but forgotten today, giving, among other things, a short chapter on charms against damage to the hives. This is an interesting survival of the old Greek custom of placing a statue of the god Pan in the bee-yard, to guard the bees from injury, a practice which we here see changed to the mediæval usage of charms and amulets.

At the end of an old French translation of the *Geoponica,* made in Paris in 1550 by Anthoine Pierre, the author in his *Salut au Lecteur,* mentions Mago (by some scholars considered one of the original sources of the book), that famous Carthaginian who lived in Libya about 600 B.C., whose twenty-eight books on agriculture, written in the Punic language, were so renowned that when the Romans captured the city these books were brought back to Rome and the Senate paid to have them translated into Latin. Later authorities have said they were "the richest booty brought back from the conquered city." The early writers all revered and followed Mago, whose works — except for a few shrewd maxims, including one or two on bees — are now lost.

Crescentius, who wrote at Bologna at the end of the thirteenth century a book in Latin, *Opus Ruralium Commodorum,* first printed in 1471, was influenced largely by Varro, Palladius and Avicenna, that great Arab philosopher who carried on the learning of Aristotle and made so great an advance in medical studies and anatomy that his works were studied at Louvain and Montpellier up to 1650. He had written on the generation of the bee. Crescentius extended the material borrowed from these earlier writers, adding chapters on gardens (a few paragraphs of this section, translated into English, are given in Trigg's *Gardencraft in Europe*), and on distilling, and treated the bees at greater length than had his predecessors. The early editions of his works are almost invariably illustrated with beautiful and interesting woodcuts.

Conrad Heresbach, whose four Latin books on rustic affairs, written largely under the influence of Crescentius, were published in 1570 in Holland, was the connecting link in bee literature between Italy and England. His book was translated in 1577 by Barnaby Googe, who frankly admits that he "increased" the original by adding material of his own. He tells us of the planting in his bee-garden, giving the medicinal uses of each plant, and it is interesting to see that he mentions Hieronymus Tragus, who wrote on herbs in Germany as early as 1539.

Gervais Markham, who took a good deal of his material from Googe, published a second edition of Googe's translation of Heresbach, in 1631, with additions of his own, changing the title; but as his own volumes — because of their great popularity — followed one another in rapid succession, he gradually substituted more and more of the fashionable pursuits of Elizabethan times, such as falconry, cock-fighting, and horse management, for the more personal matter which had given charm to Googe's pages.

Turning back to Fitzherbert and Tusser, whom Googe so admired, and who were little influenced by continental writers, we find, in Tusser particularly, a fresh, typically English country note. In both, the writing is founded on experience rather than tradition. Fitzherbert gives only a little space to bees in his *Husbandry* — the first published work on agriculture in England — but his directions for handling the swarms are interesting, as showing how it was done in 1523.

Thomas Tusser, "poet, farmer, chorister, vagabond," in 1562 had written *Five Hundred Points of Good Husbandry,* which enjoyed such a widespread reputation that in 1723 it was recommended to be taught in the schools. Written at his farm in Essex, it contained directions as to what was to be done on a farm in every month of the year, and minute instructions for the regulation of domestic affairs in general. "The best donge for the land is the master's foot walking over it," is his version of the old proverb which was known all over Europe in various forms, and even appeared later in America in *Poor Richard's Almanac*, as "The eye of the master will do more work than will his hands."

All Tusser's precepts were written in short, rhyming quatrains, full of pithy advice; and, although so celebrated because of his shrewd practical rules for successful farming, Tusser died in London a poor man, leaving nothing in his will except the money that was owed him, but having achieved so great a reputation that he was called "the English Columella."

In *May's Husbandry:*
"Take heed to thy bees, that are ready to swarm,
the loss thereof now, is a crown's worth of harm;
Let skilfull be ready, and diligence seene,
lest being too careless, thou losest thy beene."

In *September's Husbandry:*
"Now burn up the bees, that ye mind for to drive,
at Midsummer drive them, and save them alive;
Place hive in good aier, set southly and warm,
and take in due season, wax, honey and swarm.

Set hive on a plank, not too low by the ground,
where herb with the flowers may compass it round;
And boards to defend it from north and north-east,
from showers and rubbish, from vermin and beast."

In *December's Husbandry:*
"Go look to thy bees, if the hive be too light,
set water and honey, with rosemary dight;
Which set in dish full of sticks in the hive,
from danger of famine ye save them alive."

A few years later, in 1582, Batman's translation of Bartolomæus Anglicus's *Liber de proprietatibus rerum*, appeared in London. This book, which had once before been translated into English, in 1535 by J. Trevisa, the so-called Berthelet edition, was written in France in the fifteenth century and was a sort of universal encyclopædia well known in Shakespeare's time, and it must surely have been one of the books he was familiar with. Bartolomæus writes very fully of "the Bee, that little short insect" who "among all flyes with round bodyes and so shapen, beareth the price (prize)," and whose "hugenesse of wit rewardeth him in littlenesse of body."

Lyly, in his romance *Euphues,* a name he borrowed from Plato's *Republic,* gives a long description of the economy of the bee-hive, similar, but inferior, to that of Plato. This comparison of a well-ordered state to the monarchy of the bees was no new idea in literature. Cicero, too, in his "De Republica," as Shakespeare in "Henry V," had made use of the same comparison.

Numerous references of a similar nature are made to bees in the old books on emblems and symbols, so popular in the seventeenth century. The favorite emblem of Ariosto, adopted by him in 1583 when, after years of service, he was dismissed by Cardinal D'Este, was a hive of bees being consumed by flames, with the motto above it: "PRO BONO, MALUM." This was a forerunner to Shakespeare's:

> "Like the bee, culling from every flower
>   The virtuous sweets,
> Our thighs packed with wax, our mouths with honey,
> We bring it to the hive, and, like the bees,
> Are murdered for our pains."

After 1600, books on bees began to be published quite widely, appearing no longer under the wing of husbandry, but as separate volumes. De Montfort, writing in 1649, estimated that up to that date between five and six hundred authors, counting the ancients, had written on the subject.

Many of the eighteenth century bee-masters speak of Mme. Vicat, who lived at Lausanne, Switzerland, and wrote a book on bees, *Memoire pour Servir à l'Histoire Naturelle.* She was evidently a woman of alert and strong personality, and her work was highly respected by the greatest naturalists of the time, many of whom corresponded with her and placed great weight on the observations she made and recorded in her little garden on Lake Geneva. She invented a hive that was a great improvement on those which had been made before, a picture and drawings of which are given in Huisch's book on bees, along with Huber's and

Bee-house with magpie in cage

Woodcut by Hans Weiditz — German translation of Crescentius, Frankfort, 1583.

Dzierzon's hives. Mme. Vicat spent hours working in her garden, keeping her eyes on the bees and using her scissors or anything else that came handy, to aid her in her experiments. She was very successful in studying and treating their diseases, and very ingenious in counteracting the ravages of their enemies, especially the wax-moth, that enemy of the hive since the days of Virgil.

In France there has always been a great interest in bee culture, especially in the eighteenth century, when, with the great number of court functions and " polite assemblies," enormous numbers of wax candles were needed for illumination, as well as for the great festivals of the Roman church. We get some very interesting glimpses of early bee-keeping in the sixteenth century in Normandy, in Mrs. Katherine Fedden's book, *Manor Life in Old France.* This book is a scholarly study of the manuscript journal of the Sire de Gouberville, a country gentleman who lived in Normandy the early part of the sixteenth century. It is a sort of day-book, in which are recorded the everyday doings of his household.

In 1563, we find a reference to his " jardin à mouche," where, when it rained, the hives " were covered with shallow earthenware pans, made for the purpose by the potter at Saulsemesnil, and in the winter they were warmly shrouded against the cold." The strained honey was set away in " honey pots made by the potter at Brix," and, once sealed, it kept perfectly.

Another entry, in May and June, 1553, tells of several trips taken by the Sire de Gouberville to neighboring villages to buy swarms of bees. These bees were in every case bought from priests, and it is probable, as Mrs. Fedden suggests, the country curates kept bees in order to have a supply of wax for the candles on their village altars.

France has, since the time of Louis XIV, done much to encourage local industries, and in 1757, in Rouen, " for the relief of the poor and to encourage them in their labors," those inhabitants of the country who kept hives of bees were granted a diminution of their tax proportioned to the number of their hives. The French had always imported great quantities of honey and wax from North Africa, more particularly when, as happened from time to time, the bee industry of their own country languished. For these lapses, the old law " Du Droit Aboilage," was chiefly responsible. This established the right that the feudal lord, the Seigneur Châtelain, had in many parts of France to alone take the swarms of bees found in the forest in the dependencies of his seigneurie. It also gave the privilege (in many places established by custom) by which the seigneur had the right, among his other " Droits," to take a certain quantity of bees, wax and honey from the bee hives of his subjects.

We come now to the end of that long period when men knew nothing beyond what their eyes, and their ears, however keen, had been able to tell them.

Four great scientists, one after the other, were to devote long periods in their lives to the study of the bee. The first was a physician, Jean Swammerdam, who, early in the seventeenth century in Amsterdam, began to study the history of insects, working with a primitive microscope. He has been called " the Galileo of the infinitely little," because of the great new world he opened up in his book on insects, *Biblia Naturæ*. In this book, published in 1738 after his death, he made known to the world many of his discoveries, — the result of his dissection of the adult bee, the larva and the egg, and the discovery that the ruler of the hive is not a king, but a queen and mother.

The next great scientific advance was made by a Frenchman, Réamur, who was born at New Rochelle, in France, in 1683. He worked out the theory of the swarm, the life of the queen bee, and described the architecture of the hive with a care and precision which have never been surpassed. He was aided in his researches by an Italian astronomer, Maraldi, who furnished data and other material of the greatest practical assistance to him in his studies.

A third great scientist, whose discoveries were especially extraordinary because of the fact that they were made when he was entirely blind, was François Huber, who worked with the aid of a French servant whom he had trained, François Burnes, and that of his wife, Marie Aimée Lullin. This servant had remarkable powers of observation, and reported to Huber so accurately what he saw in Huber's little bee-yard, and his wife recorded so punctiliously her husband's deductions and comments, that he was able definitely to settle many of the problems that for centuries had been matters of conjecture and futile speculation. He revealed the "fairy tale of the nuptial flight, the sex of the workers, the mutual hatred of the queens, the mystery of the exuding wax, and the unimaginable architecture of the bees' communities."

The fourth man in this group of great scientists was Johann Dzierzon, who, spurred on by the progress already made, succeeded in making further successful investigations, especially in the functions of the queen. He was born in Upper Silesia, in 1811, and devoted fifty years of his life to the study of bees. He invented a hive with a removable frame which was later selected by Langstroth as the most eligible for further improvement, and so served as a stepping-stone toward the perfected American hive of 1852.

These four notebook bee-yards, where each day some new painstaking experiment was made and where new records were jotted down, are important because the discoveries made in them have entirely changed the pattern of bee life. This had hitherto gone on for thousands of years, in a routine way. From now on, a gradual change began to show itself, and new substitutes were introduced to replace the age-old instinctive activities of the bee. The modern demand for the most expeditious, the least wasteful methods of making honey has caused the bees' habits, both in and out of the hive, to be remolded on a new and more prosaic pattern; the very life of the bee has been commercialized, and the charm and romance of the old bee-yards have all but vanished.

This skeleton of bee literature, the classic tradition as it entered England, the scientific study as it progressed on the continent, is merely a paragraph or two in the annals of the bees. Were one to pursue the slender thread of bee-allusion from the time of Hesiod in 800 B. C. in his *Works and Days,* to Robert Bridges' *Testament of Beauty* of 1930, he could, following devious trails, leave hardly a corner of European letters unexplored. This rich literature — our indoor heritage from the bee — is as great a gift as that they have given us in our gardens, where the musical murmur of their wings is as much a part of the garden's magic as are the butterflies and birds, or the color and fragrance of the flowers themselves.

Bee-hives
Pietro Crescentio "De Re Rustica"
(Italian Edition) Translated by Francesco Sansovino. Venice, 1561.

HERBS FOR MY LADY'S TOILET

MRS. A. L. P. DENNIS
BERNEVAL LE GRAND, FRANCE

IN earlier days the herb bath was taken as a relief for "edgy" nerves and tired muscles, and surely we have plenty of both in these days. An old French book entitled *Mon Docteur* gives the following directions for this bath.

Leaves of Peppermint	60 grams
Leaves of Sage	60 "
Leaves of Rosemary	60 "
Leaves of Thyme	60 "
Flowers of Camomile	70 "

These are to be dried, sewed up in bags of cheesecloth, or fine linen, and a bag dropped into the bath water. *Mon Docteur* does not tell us how much to put into a bag, but I make them about three inches square and fill them as full as they will hold.

If one is not inclined to an herb bath, one can use an aromatic water as a rub-down after an ordinary bath. This is most useful after a swim in salt water, for it is very refreshing. For this, one uses:

Sage	60 grams
Melisse (Lemon balm)	60 "
Rosemary	30 "
Peppermint	30 "
Flowers of Lavender	30 "
Seed of Fennel	10 "
Cinnamon, broken fine	18 "

These leaves can be used either green or dried: the water is usually made when the fennel seeds are ripe. Put all these ingredients into a glass or stone jar and add alcohol at 90 per cent, 400 grams. Let stand for two weeks, corked tight, then filter and add 1,000 grams of pure water. It is really astonishing to see how this will bring up one's spirits after a tiresome day.

Then for an eye wash after motoring all day in the sun try an extract of Camomile flowers. This is made by putting into a jar 100 grams of the Camomile blossoms with 300 grams of alcohol at 90 per cent and 300 grams of water. This, like most of these waters, must be tightly corked and allowed to stand for a week or more when it is ready to be pressed out and filtered. Elderberry flowers dried and made into an infusion or tea also make a very good wash for tired eyes.

The Carmelite nuns used to make — perhaps they still do — an aromatic water which, if used as a rub-down before retiring, will relax one's muscles and, after a strenuous day of gardening, will help a good deal toward a comfortable night's sleep. For this Eau des Carmes, or Eau de Melisse, we should have the fresh green leaves of Melisse or Lemon Balm.

Leaves of Melisse	70 grams
Nutmeg—grated	30 "
The yellow part of lemon peel	60 "
Cinnamon, broken	15 "
Cloves—whole	15 "

Pound these well in a mortar or wooden bowl and add alcohol at 68 per cent, 1,000 grams. Let it stand eight days and filter. This is also excellent to rub gently on the forehead in case of headache.

Nowadays every woman seems bent on reducing her weight. Here is a simple remedy for this purpose, given in old herbal of 1585: "To make one slender take fennel seed and seethe it in water, a very good quantity, and wring out the juice thereof when it is sod, and drink it first and last, and it shall swage either him or her." This fennel tea is not disagreeable to take and it might help.

With the modern methods of treating the hair, with permanent waves and frequent hot oil shampoos, we are all looking for some remedy which will prevent us from losing what hair is left us. In *The Toilet of Flora,* under date of 1775, we find this: "Powder your hair with powdered parsley seed three nights every year and the hair will never fall off."

Here, then, are a few ways in which we can use some of the simple herbs which now grow in almost every garden. Of course there are creams and pomades and bleaching creams to be made of other herbs but their manufacture is a complicated business and we need not go into it now. These recipes call only for the ordinary herbs and are simple to make. Try these and see if they are not helpful.

DYEING WITH HERBS

FRANCES T. NORTON

GIVING color to fabrics through the dyer's art is not so well known a use for herbs as the blending of sweet scents and healing lotions, or seasonings and flavors for the housewife's table. It will be found, however, that it is fully as interesting to study.

Practically all the herb plants when boiled in water will give infusions which impart color to textile fibres such as cotton, linen, silk, or wool. The infusions vary greatly in strength and color with the kind of herb, the part of the plant used and the time of gathering. They also vary in the readiness with which they are absorbed into the textile fibres and in the degree of permanence or fastness of the resultant colors.

It has been a common practice from earliest times to soak into the fibre, before treating it with a dye, some solution, usually of a mineral salt which will combine with the dye in the body of the fibre to give a strong and permanent color. These substances are called mordants and in many cases, if they were not used, the dye which soaked in with water would soak out again with more water. In other words, the mordant sets the dye in the fibre.

Some of the mordants not only fix the dyes but give to them characteristic colors. Most common of these are the salts of tin and iron. Alum, the most widely used of all mordants, has, in most cases, little effect on the colors of the dye. The natural chemical compounds contained in the vegetable dyes may be profoundly changed in color by the addition of small amounts of the common and better known acids and alkalies. The well-known change of blue litmus to red when it comes in contact with acid and its turning back to blue on the addition of ammonia is typical of the behavior of many of the herb dyes.

It is an interesting experiment to fill several test tubes with the dye liquor, adding a drop of acid to one, ammonia to another, and such salts as copperas and tin chloride to others and note the change of color in the solution. This will suggest what a wide variety of colors, unfortunately not all useful or permanent, might be hoped for from the considerable number of herbs growing in our gardens and fields.

The several kinds of fibres behave very differently in the dye solutions; for instance, cotton, silk, and wool boiled in the same dye liquor will usually differ appreciably in the readiness with which they take up the dye, and will not be of the same shade of color or brilliancy.

Wool yarn seems to be the most satisfactory medium for experimental dyeing. For convenience, the yarn may be wound into uniform skeins of a practical size for small pots of dye.

Since the wool must, in most cases, be mordanted before dyeing, a brief description of the process follows.

Alum, the commonest mordant, is dissolved in water to which cream of tartar is usually added in the proportion of four ounces of alum to one ounce of cream of tartar to one pound of wool. Enough solution should be used to completely immerse the amount of wool being mordanted, and this should carry enough alum and cream of tartar to bear the proper ratio to the weight of wool used. The wool, thoroughly washed, is entered in the mordant solution at about 100° F., heated to boiling point and simmered for an hour, then drained, squeezed gently, and rolled in a linen cloth to stay for several days when, after rinsing in water, it is ready for the dye pot.

The tin (stannous chloride) and iron (ferrous sulphate) or copperas mordants are usually added to the dye in addition to the alum mordant toward the end of the dyeing process. The yarn should be lifted from the dye pot and the tin or copperas thoroughly dissolved and stirred in the dye and the yarn returned to finish the dyeing. The tin mordant tends to brighten, and the iron to darken and strengthen colors. They should be used in the proportion of

½ oz. tin crystals to 2 oz. of cream of tartar
½ oz. copperas to 1 oz. of cream of tartar } to 1 lb. wool

The dye liquor is made from different parts of the several plants, blossoms, leaves, stems, berries, fruit, bark or roots boiled for a sufficient time in water to extract their color. In most cases the blossoms yield their color most readily but do not as a rule give as permanent a dye.

The plant material should be first finely chopped, added to a pot of cold water and brought slowly to a boil and simmered for one to three hours, according to the part of the plant used. When the color is strong enough the bath should be cooled to about 100° F. and the dampened wool entered and boiled slowly until the desired shade is reached. This usually takes from a half to two hours. It is left in the dye liquor until cold. Next it is drained and washed in cold running water until no color comes from the wool. A thorough washing in hot soap suds follows, after which the wool is hung up and when dry twisted into skeins. In this manner it has been found possible, this summer, to prepare from the common herbs, nearly a hundred skeins of attractive shades and tints, most of which appear to have at least a reasonable degree of permanency.

The equipment needed for home dyeing is plenty of young fresh plant material, a few agateware pots, one to be used solely for the alum mordant, some china spoons, a pair of scales, a stove, and plenty of running water. The list of supplies should include the following:

alum	ferrous sulphide
ammonia	stannous chloride
acetic acid	cream of tartar
citric acid	soap

To this must be added a generous supply of patience. A few colors which may be made with some of the common herbs:

BEDSTRAW (*Gallium boreale*) young roots with alum, red; deeper red with the addition of tin; long boiling, terra cotta

BARBERRY (*Berberis vulgaris*) stem and roots, yellow

GOLDENROD (*Solidago*) blossoms with alum, lemon yellow; chrome yellow with tin; brown with copperas

ST. JOHN'S WORT (*Hypericum perforatum*) flowers, yellow with alum; pinkish purple with acid; brown with copperas

BLACK ALDER (*Ilex verticillatus*) bark, dark green with copperas

Hyssop (*Hyssopus officinalis*) leaves, grey green with copperas

Tansy (*Tanacetum vulgare*) young shoots, greenish grey, alum

Sweet Cicely (*Myrrhis odorata*) stems and leaves, green, alum; olive green with ammonia added

Yarrow (*Achillea millefolium*) blossoms, yellow, alum

Pokeberry (*Phytolacca*) blossoms, crimson with alum

Lichens (*Ramalina scapulorum*) red brown with no mordant; grey brown with copperas

One of the most attractive things about herb dyeing is that there is small chance of its becoming commercially important. It is quite possible to go no further than your garden to find material for an infinite variety of shades suitable for dyeing enough woolen yarn for your own use, for such work as needlepoint or knitting.

Its charm lies in the simplicity of the work and in the pleasure which comes from the development of color from common and unpretentious things.

Bibliography

Violetta Thurston: *The Use of Vegetable Dyes for Beginners.* The Dryad Press, Leicester, 1930.

Ethel M. Mairet: *Vegetable Dyes.* St. Dominic's Press, Ditching, Hassocks, Sussex, 1916.

James Haigh: *The Dyer's Assistant in the Art of Dyeing Wool and Woolen Goods.* Poughkeepsie, P. and J. Potter, Printers, 1813.

Charles F. Millspaugh, M.D.: *American Medicinal Plants,* 1887.

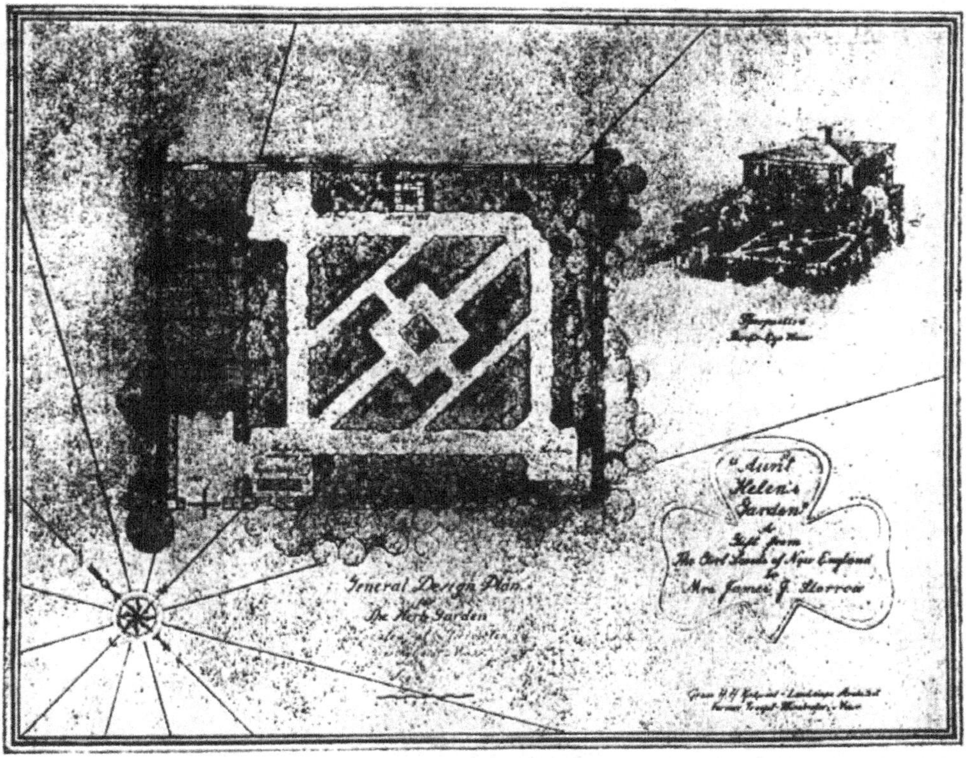

The Herb Garden
of the Girl Scouts of New England

A gift to Mrs. James J. Storrow

THE garden, placed next to the grape arbor at the rear of the Gilbert Farm House, is nearly square, being 37 by 35 feet. It is surrounded by a low stone wall, the stones for which came from Belchertown. Within the garden are flagstone walks. Close to the Gilbert House is a small mint pool, about which the several kinds of mint grow, and in which is growing watercress. The pool is supplied with fresh water by a little drip from the old stone well, over which stands a weathered well sweep. On the top of the well is an old bucket, which came from Virginia. The garden is dedicated to Mrs. Storrow in appreciation and affection for her many gifts of opportunities to Girl Scouts everywhere, as well as in New England.

More than one hundred and twenty-five different kinds of herbs were raised by the Scouts themselves or as troop projects. Mrs. Charles E. Newall, of Springfield, supervised the construction and the planting. The landscaping of the garden has been carried out under the direction of Mrs. Grace H. H. Kirkwood of Cambridge. Mrs. Hollis Webster of Lexington assisted Mrs. Kirkwood with the planting plan.

SEVEN BASILS

HELEN NOYES WEBSTER

I AM asked to write about the basils, seven of them, which were growing last summer in my garden, and which with careless familiarity I have always called by names gathered through the years from indifferent sources. And, as in search of a definite solution of uncertainties, I pursued my way through published accounts of past centuries, I found the same vagueness and confusion that, with accretions of folklore, legend and superstition, have seemingly been handed on from the old Greeks and Romans through Dioscorides, Theophrastus, and other transcribers down to the very latest of erudite garden treatises.

Now after many pleasant hours of browsing and elementary research I can truly say that I know little more than what these fragrant Ocimums in my own garden have, themselves, taught me. Spell it Ocymum if you will, but Theophrastus and others declare this to be a fodder plant, probably buckwheat, and not the herb Ocimum, so named from a Greek verb meaning " to be fragrant."

I am certain that this tropical group of plants with its comparatively few species, perhaps reducible to forty, and many horticultural and agricultural varieties and forms awaits the interest and modern revision of a systematic botanist far more competent than I. None of us who grow the tender things need to be reminded of their tropical origin. Annuals, with few exceptions, and easily raised from seed which ripens in the north none too well, the basils are mowed meekly down in their prime at the first dip of the mercury. Even heavy cloth covering will not save them for long if the days become cold. Paradoxically the seed of all our garden species sows itself if we leave the beds alone in the fall and spring. This is more easily said than done; for is there anything which can look as sorry and bedraggled as the first frozen basils when perhaps all around late thymes, hyssop and costmary are still blooming? But cover their demise with a heavy mulch of leaves and get your reward in very late spring when tiny purple and green leaved seedlings make their unexpected appearance. Of course the growing season may be long enough to give us ripened seeds to garner, packet, and label. But when you think that all is lost in premature frost, try mulching the frozen plants.

The East Indian basil, Tulasi, *Ocimum sanctum L.*, glues its falling seeds to the broad leaves below the flowering spike, spotting them as if with some fungus disease. The leaves fall and are wind-tossed away. I cannot find this fact recorded nor several other characteristics which make the Sacred Basil a good plant to grow in any kind of New England garden. It is a pretty, little, single-stemmed shrub, topped with three spikes of crowded soft purple flowers, the middle spike being always the longest. The whole plant is hairy, quite different in appearance from the basils of the O. Basilicum group, and has a curious resistance to frost which by all rights it should not exhibit. The Hoary Basil, *Ocimum canum*, although with whitish flowers, is so nearly like the Sacred Basil, *Ocimum sanctum*, that only as they grow together can I see any difference. The latter is taller, more shrubby, with lavender flowers, but both have the same fragrance of clove, lemon and cinnamon. O. canum Sims was first described in 1824 and pictured in *Curtis' Botanical Magazine*. The drawings here were made from plants raised from seed sent from China to the Directors of the Horticultural Society in England and grown in their Botanic Garden. It is closely related to *Ocimum Indicum*, a species frequently described in old herbals. Parkinson makes much of this Indian species, but it should not be confused with the species we know today as *Ocimum sanctum*. The latter is of East Indian origin, and grows in the Philippines, though Dr. Merrill says that " its appearance in these islands indicates only prehistoric introduction." Like other basils, both the Hoary and the Sacred Basil have been since early days of medicinal importance. Its precious sanctity assures entrance into the Hindu heaven and homes are guarded by the protective spirit of this herb. I once showed, to a missionary, who had known India for many years, these two species after they had reached midsummer height in my garden. " Yes," he said, touching first the leaves of Hoary Basil, and then Tulasi, " that is the most sacred plant in India." Some day I hope to grow, with these two basils, shrubs of the Tree Basil, *Ocimum gratissimum, Linn.*, which is said to be decidedly ornamental. The long heavy spikes of flowers are not unlike those of our wild Clethra. It is the " basilic en arbre " of the French kitchen garden.

Ocimum suave, Willd., the fever bush from Abyssinia, with its broad woolly dentate leaves and long spiked, close-set flowers, would fascinate me both by its beauty and long medicinal history in the welfare of the natives.

Concerning all the basils there is pleasant humor in the exuberance of exaltation or intolerance with which the ancient authors proclaim or denounce the virtues of this herb. The very earliest reference, to my knowledge, is where Chrysippus, about 250 B. C., declares that "Ocimum exists only to drive men insane." When Greeks heaped curses on its leaves, to most Romans it was the herb of love. " Even," says Cassianus, " the malignant quality of the herb is indicated by the fact that a she-goat that browses on everything, avoids Ocimum alone." And no ancient description is complete without direful warnings of its scorpion-breeding tendencies. But English literature teems with delightful allusions to basil. Read any of the Tusser editions of *Five Hundred Points of Husbandry* since 1557; turn to his directions for May and then go stroke your basil, as he bids, that " It may grow and multiply " the better for your sympathy.

Do we read with amusement of those superstitions which ring so true of belief? Do we remember how closely they are connected with all the herbs that have come from the Old World into the gardens of the New? Almost as precious as the Bible to the early settlers were their herbals of Tusser, Culpepper, and Coles, which contained all that was necessary for the housewife to know about the herbs, their uses and the art of simpling. Sweet basil is one of the herbs mentioned before 1806 in colonial garden records. Then we read of South American basils finding their way into European gardens.

Just when the mosquito bush, or wild basil, *Ocimum micranthum, Willd.*, was brought from Yucatan into Florida is uncertain, but it was so very long ago that Gray listed it in his " Synoptical Flora " as native. It seems likely that this basil is the same species as *Ocimum Campechia-*

num which Miller describes in his *Dictionary of Gardening* (1759) and calls the "Sweetest basil of Campeachy." Seeds were sent at about this time by Dr. Will Houston to England, and if we inquire as to the whereabouts of Campêche we learn that it is in Yucatan. There was probably a plant of *Ocimum micranthum*, or the Campêche basil, in my garden last summer; at least it seemed to have a well-labelled authentic counterpart among the pressed plants of the Gray Herbarium. This basil was very different from any others in my acquaintance. It is a low bushy plant, nearly smooth, with long petioles, ovate leaves, and white flowers, which, like the foliage, shows slight tinges of purple. I never had enough leaves for culinary experimentation, but, as Small's *Southern Flora* lists this basil as not uncommon in Florida hammocks, we may leave that to our southern members. The odor is flowerlike rather than spicy.

How valued an herb was basil is proved by the active exchange of its seeds between countries for many, many years. Then there was a long period of apparent apathy. War-torn states in the middle of the last century were not concerned with foreign plant introductions unless of importance as herbs of healing. Now once more there is interest in gardens of sentiment and fragrance. Lacking its basils, no herb garden is perfect. Ethiopia, Madagascar, Abyssinia, offer rich collecting grounds, and from these sources will come "age-old" forms of this historic genus as novelties to add interest to the gardens of America.

A unique species of Ocimum should in time find its way into our summer gardens. There is no more reason why the Saline Basil of Chile, Ocimum salinum, should not be one of our conservatory and garden habitants than heliotrope or fuchsia. The description of this plant in Molina's *Natural History of Chile*, as translated from the Italian by "An American Gentleman" in 1808, follows:

"In the Province of San Jago in Chile is found a species of wild basil, Ocimum Salinum, differing in its appearance from the common or garden species only in its stalk which is round and jointed, but in its smell and taste it resembles more the alga or seaweed than the basil. This plant continues to increase in growth from the first opening in spring to commencement of winter, and is every evening covered with saline globules that are hard and shiny and give it the appearance of being coated with dew. The husbandmen collect and make use of this salt instead of the common kind which it far exceeds in taste. Each plant produces daily about one half an ounce, a phenomenon the cause of which I am not able satisfactorily to explain as it grows in very fertile soil, exhibiting no appearance of salt, and is more than sixty miles from the sea."

Miller speaks of the "Studded Basil" as one of the varieties of "Great Basil," *Ocimum Basilicum*, and the ardor with which those seventeenth century horticulturists collected from every corner of the world arouses the surmise that this may be the Chilean Basil, although Miller does not speak of the salt.

To pass on to the well-known species and varieties of basils in the modern garden, we find, from *The Gardener's Dictionary*, written by Philip Miller, who was gardener to the Royal Society of Apothecaries in the garden of Chelsea (1768), that the growers of basil two centuries ago were assailed by the same perplexities as are our nurserymen of today. Note this comforting comment. Miller speaks of three "sortes of basils" growing naturally in India and Persia. "Of them there is a great variety which differs in size and shape and color of their leaves also in their odor, but as these differences are accidental so I have not enumerated them being convinced from repeated experiments that the seeds of one plant will produce many varieties."

Would that I had one of those prize packets! There may be some, however, who in their dealings with uncertain nurseries will remark caustically that it is not less so today. To read Miller's picturesque description of Great Basil's varieties is to be grateful that Linnæus was still in the throes of binomial nomenclature. His *Ocimum Basilicum*, sometimes called the "Great Clove-scented," is allowed five varieties, and we can almost see them growing:

1. Fringed-leaved basil, with the purple leaves
2. Green-fringed basil
3. Green basil, with studded leaves
4. Large-leaved basil

What more has Vilmorin to offer in his 1935 catalogue? He lists, under Basil, "Tall Green; Tall Violet; Delicate Green, suitable for stews; and Lettuce-leaved."

Before 1848, when De Candolle made his long list of Ocimum species and varieties from all over the world, the early herbalists seemed to content themselves with these three "Sortes," the Great Basil, the Medium

Ocimum canum (Hoary Basil)
From Curtis Botanical Magazine, *Volume* 51, 1824.

Basil, and the Lesser Basil, and this naïve classification is common in herbals up to the eighteenth century.

Exactly to what modern species some of these might be compared we can only conjecture. Certain it is that the Great Basil is *Ocimum Basilicum L.*, the kingly herb, royal enough to be carried by royal hands into royal chambers," and that no herb is more well defined in color, flavor, and habit. The yellow-green bushes a " cubit high " give welcome garden color for us between summer savory and wild marjoram. Its white flowers follow up in their opening the leafy stem for several weeks. This makes it one of my best bee plants. The leaves fold on their midribs in extremes of heat and cold, and their flavor is indispensable in cookery. The southern races know well the zest given by a few shreds of leaves or flowers sprinkled over fish, tomato, or spaghetti. When green, the leaves make a pleasing oil, and when dried are used as a snuff cure for nervous headaches. Like all other basils, it grows easily and quickly from seed sown when the earth is warm in well-nourished garden loam.

I greatly doubt if as such the seeds of Curly Basil may be bought in this country, but this basil is always appearing in our collections. It is probably a French agricultural variety of so-called Lettuce Leaf, but it is well worth growing under any name. The shrubs are coarser and over-top all my other basils. Its large wrinkled leaves and short spikes of crowded flowers make, when dried, the best herb pepper when in combination with costmary, savory, and marjoram. The fragrance may not be as pleasing but it is powerful and resinous.

This last form is sometimes spoken of as belonging to the " Medium Sorte " of basil. Perusal of old descriptions and study of woodcuts leave me nowhere as to its identity. I'm sure it must be among my garden basils, for after reading the list of varieties which Miller accords to Medium Basil I really don't see how we can miss it.

1. Common Basil with very dark green leaves and violet flowers.
2. Curled leaf Basil with short flower spikes.
3. Narrow leaf Basil smelling like Fennel.
4. Middle Basil with citron scent.
5. Basil with leaves of three colors.

What if imaginary aromatic differences are more marked than structural? What if next year each may shade off into something else — thus escaping binomial designation? I greet them every spring with friendly recognition, even as did the eighteenth century apothecaries' gardener.

Have I left purposely to the last the third " sorte," the little Lesser Basil? How easily we fall into the picturesque phraseology of those early herbalists!

Bush Basil, *Ocimum minimum, Linn.*, is as sharply defined a species as its taller cousin, Sweet Basil, and equally important in home, garden, and legend. Its small spicy bushes are so round and compact that they might have been treated by a topiarist. It was a favorite herb for pots with the English cottager who often gave it in compliment and well-wishing to a departing guest. The tiny green leaves are smooth and thick set on the many branches, and the white flowers are in leafy little spikes. Its two varieties I know as " Sweet Purple Bush," a name from our Southern gardens. They must be what Miller describes as " the smallest basil with black leaves and the smallest basil with variable leaves." The tiny purple flowers are always humming with honey bees which bend in their zeal the little six-inch bushes to the ground. Their fragrance combines all the odors of eastern spices. Do its leaves like those of Green Basil cure deafness when laid in the ears, make sweeter washing waters, or more potent medicinal oils? All this I do not know, but in sweet nosegays it excels.

And if of my basils I have talked too much and of more than the allotted seven, I offer no apology, but the assurance that if I am privileged to share any of them with my readers they will do likewise.

SHAKESPEARE'S FLOWERS

ELIZABETH WADE WHITE

"But thy eternall Sommer shall not fade."
—SONNET XVIII.

CAN the poet who, more than any other human being in the world's history, is "all things to all men," of whom St. Beuve said that he was "*la nature même,*" whose miraculous catholicity of knowledge and experience is an unending source of wonder, be said to belong, to any degree, more to one class or group of mankind than to all the others? Yes, Shakespeare is predominantly the gardener's and countryman's poet. A scion of the English yeomanry, brought up in those pleasant meadows of the Avon from which his life was never too far removed for the final return to their tranquillity, he absorbed with the breath of his youth every minutest detail of their existence, and never lost the instinct to distill those country impressions into the essence of his poetry. Lover and beloved of the kaleidoscopic city of London, friend of starving poets, quarreling actors and bejeweled courtiers, and so infinite in his understanding of the violence and beauty of the man-made world, he was always, and to a far greater extent than his contemporaries, aware of the vast and immortal background of nature. This preoccupation with country things has been dealt with but recently, and I presume more thoroughly than ever before, in a scholarly and fascinating book: *Shakespeare's Imagery*, by Professor Caroline Spurgeon, and it is impossible to resist the temptation to quote the following passage, in order to usher in the discussion of another and kindred book* with, as it were, a confirmatory flourish of trumpets and hautboys:

"One interest, above all others, stands out in Shakespeare's imagery. This is the life of the country-side and its varying aspects: the winds,

**Shakespeare's Wild Flowers, Fairy Lore, Gardens, Herbs, Gatherers of Simples, and Bee Lore,* by Eleanour Sinclair Rohde, London, The Medici Society, 1935. 8s. 6d.

the weather and seasons, the sky and clouds, birds and animals. One occupation, one point of view, above all others, is naturally his, that of a gardener; watching, preserving, tending and caring for growing things, especially flowers and fruit."

And here we have — as Professor Spurgeon's book is written for students of Shakespeare the man and writer — a new and charming book for those lovers of gardens and poetry who concern themselves with the old English flowers and with the poet who has preserved them most freshly and vividly for us: *Shakespeare's Wild Flowers, Fairy Lore, Gardens, Herbs, Gatherers of Simples, and Bee Lore*, by Eleanour Sinclair Rohde. And as Professor Spurgeon has focussed new light on Shakespeare's life and character through a study of his general imagery, so Miss Rohde has illuminated our understanding of 16th and 17th century English garden and nature consciousness, and thus greatly enriched the store of garden literature, by her collective examination of Shakespeare's nature imagery.

Shakespeare was never a botanist, although it is conjectured that at some time during the obscure years, 1584-1592, he *may* have practiced as an apothecary. But his flowers and herbs are neither the pressed and desiccated specimens, nor the scentless scientific symbols, that the past has often handed down to us; rather they spring as vigorous and dew-brightened in his pages as they did, three hundred-odd years ago, from their native English soil and in his all-comprehending observation and memory. And Miss Rohde is well qualified to up-fill a book, as Friar Laurence did his " osier cage," with " children of divers kinds " from that most evergreen of gardens. For she has explored, with patient and scholarly imagination, very many mysterious and all-but-unvisited gardens of the past, some of them so overgrown with crabbed manuscript thistles or black-letter briars as to necessitate the costume contrived by " Sir " John Hill's herbalizing friend, which " had a Mask for his Face, and Pads to his Knees, that he might creep into the Thickets." From these expeditions Miss Rohde has brought back one delightful and informative book after another, but never has she offered us a more fragrant harvest than in the present volume.

There is an ingenious continuity in the sorting, arranging, and presentation of this wealth of quotation and allusion. We begin, simply enough, in an English wood, in the spring, with wild flowers, " pale primroses," " the freckled cowslip," " violets dim," about our feet, and English oaks above our heads, and so through Shakespeare's varied store of native plants and trees to a chapter on the oldest inhabitants of those woods, the company of Faery. These elusive, fascinating, and often sinister creatures, so keenly sensed by each successive generation of mankind, and yet, even in Chaucer's time, so rarely seen, are really dependent for their existence on the poetic imagination, and no age has given us a more enchanting race than the Elizabethan fairies, nor a more lovely fairy queen than Shakespeare's own Titania. And so to the world of man, with its ordered and tended plantations of flowers and fruits, and to *King Richard the Second,* that sombre play with its underlying symbolism of a neglected garden, and its now particularly poignant use of an old belief:

> " 'Tis thought the King is dead; we will not stay.
> The bay-trees in our country are all wither'd."

Shakespeare loved roses, lilies, and their stately companions in the manor-gardens and pleasaunces of the great, and he knew, with perhaps even a more familiar affection, the simpler gardens of town and village, where flowers, herbs, and fruit grew together in aromatic congeniality. His knowledge of horticulture, of orchard and vineyard tending, and of the care of bees, has an undeniably first-hand flavor, and not one of his contemporaries, save the herbalists themselves, spoke with such frequency and acquaintance of herbs and their preparation. Miss Rohde's chapter on his " Herbs and Gatherers of Simples " is fascinating, as is also the discussion of his bee lore, with its glimpses of the " King-Bee " and of the many Elizabethan uses for honey and wax; and so we come to the last chapter, " Suggestions for Making a Shakespeare Garden." This is indeed a valuable and stimulating contribution by one who has perhaps the most creative influence of any modern garden writer on those modern gardeners who wish to use both tradition and imagination in the making of their gardens.

Miss Rohde has added another mellow and well-constructed book to the already long list of her previous publications.

ON BEHALF OF PERFUME

FREDERIC WARDE

PERFUME is one of the elements and one of the secrets of the universe. The history of perfume is the history of mankind from the beginning. This history has been written more completely in perfume than in blood and tears. It antedates the Christian era by many thousands of years. The Chinese made perfumery one of the accessories of their civilization. The Egyptians used perfumes as offerings to their deities and kings. Old incense indestructibly still hovers about ruined shrines, and our imagination may re-create for us the magnificence of great kings whose sarcophagi still exhale frankincense, myrrh and other aromatics.

The literature of all nations and ages is permeated with references to perfume. The Old Testament is a rich source: The first Divine commands given to Moses (in the Book of Exodus) were directions for erecting an altar of incense, and for compounding the holy oil and perfume to be used in the anointing of all sacred things besides the heads of kings. In the *Iliad* there is a description of the toilet of Juno. The Elysian Fields, the Greek's idea of Heaven, were made of perfumes. In the Elysian Fields there was a golden city with ramparts of emeralds and gates of cinnamon. About the city flowed a river of perfumes, one hundred cubits wide and deep enough to swim in, from which an odorous mist rose and enveloped the whole place and then descended as a refreshing and fragrant dew. In this city, too, there were hundreds of fountains of honey and of the sweetest essences. The ancients thought of perfume as one of the attributes of the gods and goddesses. The first perfume came from Venus when her nymph, Ænone, fell in love with a mortal; and, thus, it became the treasure of kings. The immemorial association of perfume with beautiful women may be one of the divinities which Plato declared as always residing in beauty.

The first perfumers were the priests of the earliest religions. The making of perfumes was one of their duties and one of their mysteries. The Greek poets and dramatists have given us the names of several famous perfumers. Old Megallus and a fashionable perfumer, Peron, of Athens, the violet-crowned city, still live in an old play. One of the earliest writers on the subject of perfumery was Theophrastus, a Greek and a pupil of Aristotle, who died in 278 B. C. A translation of Theophrastus' interesting treatise may be found in the Loeb Classical Library. Virgil, Pliny the Elder, and many other Latin writers, have given to us their accounts of sweet aromatics. One writer relates that, at banquets, while guests were at table, the air was filled with a fragrant, iridescent mist rained down by doves whose wings had been saturated with perfumes. The Romans made perfumes from roses, quince-blossoms, bitter almonds, and from the narcissus. Fashionable men and women of Rome made use of a different perfume for each part of their body, and for each of their garments. Horses and dogs were rubbed with scented pomades. Saffron was used to perfume military flags and the awnings on houses and over the amphitheatres.

Nothing has yet eclipsed the power of perfume to invest us with charm and distinction wherever it has left its lingering touch. To the refined and delicate side of life, perfume is affiliated to aristocratic and enlightened living.

The perfumer is an artist and a poet. He creates perfumes with the aid of science and psychology. Perfume, music, poetry, and color have strange powers over the mind. They awaken the memory and the imagination; they recall far-away events. There is no scientific proof of these effects which produce some kind of deep spiritual apprehension. How would the poets have expressed themselves without thoughts of perfume, music and color? How would Virgil have written his *Second Eclogue,* which is a poetic combination of odors and colors?

> See, for thee the nymphs bring baskets heaped with lilies: for thee the winsome Naiad, plucking pale violets and poppy-heads, weds jonquil to scented fennel-flower, and anon, enweaving them with casia and all sweet herbs, gilds the velvet hyacinth with the yellow marigold! My own hands shall gather thee quinces, gray with tender down, and chestnuts such as my Amaryllis loved. Waxen plums I will add (for plums, too, have their meed); and you, O laurels, I will pluck, and thee, their neighbouring myrtle—for set side by side ye blend your sweet fragrances. . . .

Many writers, besides Huysman, in *A Rebours*, and Huneker, in *Ivory, Apes and Peacocks,* have written about the effects of perfume.

In his book on the art of perfumery, Piesse has compared odors to sounds. "Scents," he has written, "influence the olfactory nerve in certain degrees, just as musical sounds influence the auditory nerve. As there are octaves in music, there are octaves of odors. To make a bouquet that is harmonious, the perfumer should select odors that chord together." Since the publication of the book by Piesse, many entirely new perfume bases and numerous shades of each type of odor have been discovered by chemistry. Instead of two or three hundred, the perfumer now has almost a thousand different essences which he may use in his compositions.

Fresh flowers exhale the most refreshing fragrance. In the petals of odorous blossoms there is a minute trace of an essential oil which produces the natural perfume. This oil exists in a free state in the rose, in lavender and in other blossoms. In some kinds of flowers the odor may be the result of a chemical change; in the jasmin and the tuberose, it is the result of a fermentation. Mr Poucher, in his work *Perfumes, Cosmetics and Soaps,* has given a list of the parts of plants which yield the desired oil:

FLOWERS: Cassie, Carnation, Clove, Hyacinth, Heliotrope, Mimosa, Jasmin, Jonquil, Orange-flower, Rose, Reseda, Violet, Ylang.
FLOWERS AND LEAVES: Lavender, Peppermint, Rosemary, Violet.
LEAVES AND STEMS: Cinnamon, Geranium, Patchouli, Petitgrain, Verbena.
BARKS: Cinnamon, Cassia.
WOODS: Cedar, Linaloe (Rosewood), Sandalwood.
ROOTS: Angelica, Sassafras, Vetivert.
RHIZOMES: Ginger, Orris, Calamus.
FRUITS: Bergamot, Lemon, Lime, Orange.
SEEDS: Almond (bitter), Anise (both kinds), Fennel, Nutmeg.
GUMS: Benzoin, Labdanum, Myrrh, Olibanum, Peru balsam, Storax, Tolu.

Climate, soil, and different other conditions, influence and cause a plant to produce an oil of good or bad quality.

The study of the formation of essential oil in a plant has captivated the interest of distinguished scientists. There have been many different theories. After many years devoted to experiments, Mr Charabot and Mr Laloue have shown how, in many cases, essential oil originated in the chloroplast in a plant, and was the result of the assimulative work of the chlorophyll. Information regarding the individual constituents of essential oils is the result of a science involving many other sciences. The perfumer is fortunate to have the fruits of much research in the form of an extensive gamut of odors with which he can produce scents that will transmit messages to the mind. As the poet uses words to create a certain picture, the perfumer, by his learned sensitiveness and his delicate and remarkable discrimination, does the same thing with a scent that has many different nuances — almost imperceptible to an ordinary nose.

Various processes are employed to separate oils from flowers and other substances. The processes are, distillation, expression, extraction with pomades (enfleurage with cold fats; maceration with hot fats), and volatile solvents. Distillation is the oldest, the most general and important process used to obtain essences and aromatic chemicals. There are many different kinds of alembics, retorts or stills. The quality of the essence distilled depends on the refinements and the scientific construction of the alembic used, and on the technical control of the process of distillation.

Attar of roses (*Attar-gul:* Persian for " essence of roses ") is obtained from the blossoms distilled in small or large alembics over an open fire; in large alembics heated by steam; and by volatile solvents. The Bulgarian attar of roses is from the *Rosa damascena* which is grown in Bulgaria. Before 1900 the Bulgarian cultivators distilled their own roses in small alembics over open fires. The small alembic is made of copper. An outline of the shape of this apparatus suggests the body of a short and very fat man with a large head. The body of the alembic may be large enough to hold 100 or 120 litres (25 to 30 gallons) of water. These alembics are placed in rows on a low brick hearth. When the roses have been gathered, the white and red blossoms are mixed together and distilled without delay. Distillation is completed in about two hours. The proportion of roses to the water in each alembic is one of roses to five of water — ten kilos of roses to seventy-five litres of water. The quantity of rose water collected from each distillation is ten litres. This rose water is put aside. The alembic is emptied, and the exhausted roses are discarded. The hot water is returned to the body of the apparatus; the quantity of this hot water is fifty litres; twenty-five litres of fresh water are added to the hot water with another charge of ten kilos of fresh roses. This process is repeated and continued till the day's gathering of blossoms has been distilled.

The rose water collected from four distillations, forty litres, is put into a clean alembic and distilled. The liquid of the first five litres collected from the second distillation is cloudy. This distillate is reserved and allowed to stand; when it becomes clear, the oil of roses, which collects on the surface of the water, is removed, and put into flasks which are sealed and stored. To the thirty-five litres of rose water, remaining in the body of the alembic, five litres of the rose water of the first distillation are added. This procedure with the second distillation is repeated until all the oil has been separated from the rose water. This essence is called "peasant quality." Whether small or large apparatuses are used, the proportions are exactly the same: One kilo of roses to five litres of water, to yield one litre of rose water from one kilo of roses.

The process of extraction by volatile solvents is, comparatively, a simple one. Various kinds of refined ethers are used. Petrolic ether is generally favored because it volatilizes at 25 to 60 degrees C., and leaves no residue or odor of its own. Flowers are placed on trays that fit inside large cylindrical tanks. Each tank will hold 100 gallons; and, when charged with flowers, each tank is completely sealed. Series of these tanks are connected together by various tubes which conduct the liquid ether in and out of each tank in rotation. When the solvent has saturated and exhausted the flowers of their oil, it is drained off. Fresh flowers are replaced in the tank, while fresh solvent is extracting oil from the flowers in the succeeding tank. The process is continued without interruption. The solvent drained off the flowers is run into a vacuum alembic which is much the same as any alembic, but made so the air inside its body may be reduced below atmospheric pressure. By reducing the atmospheric pressure in the alembic, the liquid can be volatilized at ordinary room temperature. In the vacuum alembic the solvent is easily volatilized, recovered, and returned to the solvent tank from which it will be withdrawn and again mixed with fresh flowers. The residue left in the alembic, after the ether has evaporated, is a thick mass that contains the natural oils, gums, and waxes, of the plant or flowers. This mass, commonly called concrete, is dissolved in strong alcohol; the alcoholic solution is then refrigerated at about 15 degrees below zero. The more or less soluble waxes will harden at this low temperature, and the liquid alcohol, retaining the precious essence of the flowers, is drained off. This alcoholic solution is put into another vacuum alembic and is distilled; the oil left in the alembic is the pure absolute of the essence of a flower.

Essential oil that has been distilled by water or by steam will be deficient in the full amount of some of the odorous constituents, because some of the constituents are soluble in water, and will be retained in the water left in the alembic. Absolute essences obtained by volatile solvents have all of their constituents in their natural proportions.

The nose is still the most reliable critic of the quality of the odor of an essence. Many elaborate tests have proved the nose is about 75 per cent reliable against 25 per cent by all the chemical methods known. Chemistry determines what constituents are in certain oils and perfumes; but the olfactory nerve remains the supreme judge of the odor. An experienced nose can detect in the odor of a sample of attar of roses whether more red or more white roses were used in the production of the oil; whether, by its soft sweetness, it is a peasant oil, or, by its intense sharpness, it is an oil from the mass production of large alembics; whether, by its aldehyde character, the oil has come from a steam alembic, or, because of its leafy heaviness, it was obtained by volatile solvents; and whether or not much time elapsed between picking and distilling the flowers. Perfumes are composed according to formulæ. A formula systematically sets forth the exact amount of each ingredient required in the composition of a particular perfume. All perfumes are syntheses: The characteristic fragrance of any one flower is the result of the formation in the flower itself, or the re-formation by the perfumer, of a compound by combination of its elements or constituents.

The most accurate formulæ have their ingredients compiled metrically. The simplest way to express the proportion of one ingredient to another, and of the total, is by percentage. It is an advantage to be able to decimate a formula, in order to reduce or increase the formulated quantity. An illustration of this advantage may be seen in a formula for a Jasmin Base —

200 grams	Benzyl alcohol	20.00	2.00
550 "	Benzyl acetate	55.00	5.50
150 "	Linalyl acetate	15.00	1.50
100 "	Linalol	10.00	1.00
1000 grams		100.00 grams	10.00 grams

The proportions remain unchanged. Formulæ that have their ingredients compiled in unrelated quantities, such as ounces, drams, grains, are bastard formulæ; the unrelated quantities should be resolved into metric proportions in grams.

Perfumes should be compounded by weight only, and by accurate weighing, because the strength of essences is determined by weight instead of by volume. A certain quantity of an essence or substance, liquid or dry, will vary, relatively, when compared to the weight of some other element that is taken as a standard. The standard weight is that of one cubic centimetre of water at four degrees Centigrade, and this is one gram.

For weighing things, there are in general use two systems, the English system and the Metric system. The English system (the grain, ounce, pound, as units in weight, and the pint, quart, gallon, as units in volume) is really not a system but a very primitive scheme; the units are not related in any way, and they have been arrived at in the most arbitrary manner imaginable. The Metric system is the simplest and the most scientific system known, and it is the only system used by chemists the world over.

1 kilogram: 1000. grams		
1 hectogram: 100. grams		
1 decagram: 10. grams	1 gram: 1 gram	1 decigram: 0.1 gram
		1 centigram: 0.01 gram
		1 milligram: 0.001 gram

Special forceps are provided with a set of weights; this instrument should be used without any exception. Weights should NEVER be touched by the fingers; the natural moisture of the fingers will affect the accuracy of a weight.

It is necessary to have a good balance that has knives of polished steel and agate knife-rests. The balance should have pans that are about eight inches in diameter. A perfumer should have a balance which is capable of accurate counterpoise with 1000 grams in each pan, and which is sensitive in reaction to 0.01 of a gram. A rider-scale up to ten grams, divided into tenths of a gram, is a useful feature. The base of the balance should have pointed levelling screws, and there should be a glass rest under the point of each screw. It is imperative to keep the balance perfectly clean, in a suitable case, where there is no heat, no current of air (even one's breath will divert the swing of a balance when one is weighing something), and where there are no vibrations. As for glass measuring cylinders, there should be one to hold 100 cc. (one litre), a second one to hold 500 cc., and a third one to hold 100 cc. A porcelain mortar that will hold two or three litres is very useful in a laboratory. Work in perfumery requires many bottles. It is economical to purchase bottles in lots of a dozen. The bottles should be made of amber glass which will protect essences from the injurious effects of light. For all solutions, tinctures, and perfume bases, and for perfume extracts being matured, one should use amber glass bottles with glass stoppers. It will be necessary to have bottles of various sizes: 15 cc., 30 cc., 60 cc., 125 cc., 250 cc., 500 cc., and 1000 cc.

Gummed labels of a suitable size will be needed, for all bottles containing compounds and solutions should be carefully labelled. Each should have the name, number of the formula, per cent in strength, and date, of the contents of each bottle. It is obvious that one keeps a record of the formulæ compounded; a loose-leaf book is good for this purpose. Put only one formula to each page, number it, date it, and add notes regarding the results achieved. A calendar should accompany the record, owing to the fact that at least thirty days must elapse before a perfume will mature; a note against the future date should be a reminder as to when a perfume may be judged.

Essences and other materials, needed for making perfumes, should be purchased from reliable manufacturers. Some manufacturers specialize in only essential oils, some others specialize in only aromatic chemicals, and a few large firms sell both oils and chemicals. Usually, there is one firm, or another, that is known for its superior quality of an oil or a chemical. The perfumer must learn to know what he wants and how and where to search for it. He will find the desired product, and he will often make discoveries which should stir his imagination. The amateur perfumer may and should purchase all of his different oils and substances in one-ounce bottles and containers. The firm, from which materials are purchased, will deliver its products in bottles which are sealed. When a bottle is opened for the first time, the cork should be removed, carefully, with a cork-puller; the removed cork will have a hole through it and should be discarded, and a new cork of the best quality should be inserted in the bottle. Beware and avoid mixing the corks that go in any of the bottles used for perfumery. The bottles that form the perfumer's arsenal may be arranged in alphabetical order on shelves in a cabinet which can be closed to light, or in a series of drawers.

In addition to the essential oils, the perfumer must have several gallons of the best quality of pure grain alcohol that is 95 per cent in strength. The alcohol must be almost odorless. For making solutions of various kinds, distilled water is necessary. Unless one has an alembic, it is well to obtain distilled water in five-gallon containers.

Before a perfume may be composed there are solutions of various kinds to be prepared. Each of the fatty aldehydes is too powerful to use it pure; these aldehydes must be reduced with phenyl ethyl alcohol to make 10 per cent solutions of each of them. To do this, counterpoise a 15 cc. bottle on the balance, and weigh into the bottle 9 grams of phenyl ethyl alcohol and then 1 gram of the aldehyde; insert the glass stopper, shake the contents, and label the bottle. The solutions required for a perfume are always indicated in the formula. The alcohol, used for making a perfume, should be deodorized by dissolving 1 gram of Benzoin resin to 1 litre of alcohol and matured for 30 days.

There are numerous aromatic waters used in the preparation of toilet waters. The United States Pharmacopœia IX gives the following instructions for making aromatic waters — "Aromatic waters, when prepared from volatile oils, are intended to be, as nearly as practicable, saturated solutions which must be clear and free from solid impurities, and, unless otherwise directed, should be made by the following process:

2 grams Volatile oil
15 grams Purified talc
Distilled water sufficient to make 1,000 cc.

Triturate, in a mortar, the volatile oil with the purified talc, add the distilled water gradually with constant trituration; filter, and pass the filtrate through the filter paper repeatedly until the aromatic water is perfectly clear. Do not let the solution freeze." Many aromatic waters can be made by this process: Anise, cinnamon, fennel, peppermint, spearmint, rose, orange-flower, and geranium.

Toilet waters contain a certain quantity of alcohol. The odor of a toilet water is derived from the perfume which has been dissolved in 95 per cent alcohol; this solution is then reduced with distilled water to the desired alcoholic strength. One should have a table of proportions for the dilution of alcohol with distilled water. The table should give both the weights and the volumes of the alcohol and the water required to obtain 1000 grams or 1000 cc. of diluted alcohol from 96 per cent down to 30 per cent. A simple method for transforming the strength of alcohol with distilled water is — if 95 per cent alcohol (or any other per cent of alcohol above 50 per cent) is to be reduced to 30, 40 or 50 per cent alcohol, take a graduated cylinder that holds 1000 cc., fill it to the 300, 400, or 500 cc. mark with the 95 per cent alcohol, then add distilled water to 950 cc., or to the amount that corresponds with the per cent of the alcohol used. If the alcohol on hand is only 70 per cent alcohol, fill the 1000 cc. cylinder to 500 cc. and add the distilled water to 700 cc., to produce a solution that is 50 per cent in alcoholic strength. Most toilet waters contain alcohol which will be from 50 up to 70 per cent in strength.

An eminent French perfumer, Henri Fouquet, recommends an excellent process for making toilet waters:

5.00 grams Perfume
2.50 grams Magnesium carbonate
2.50 grams Glycerin
1.25 grams Borax
Alcohol, 60 per cent, to make 1000 cc.

1. Select a bottle that holds 1500 to 2000 cc. Pour the 1000 cc. of Alcohol into the bottle.
2. Select the essence, the perfume base—not an alcoholic extract. Weigh the 5 grams of the Perfume into a glass watch crystal (4 or 5 inches in diameter), add the 2.5 grams of Glycerin.
3. Triturate the Perfume and the Glycerin, add the Borax, continue the trituration until the mixture becomes a liquid; add the Magnesium carbonate, and triturate until the whole mixture becomes a gummy mass.
4. From the 1000 cc. of Alcohol, add enough Alcohol to thin the mass so it can be poured into the bottle which contains the Alcohol. Continue to rinse the mortar until all the substance has been transferred to the Alcohol.
5. Insert a new cork in the bottle, and shake the contents for thirty minutes. Agitate the mixture for five minutes each day for a week. Keep in a cool place for one month. Filter through fine filter paper until the liquid is perfectly clear. The toilet water will be delightfully perfumed, and the liquid will feel soft and grateful to the skin.

The formulation of a perfume is divided into six parts: 1. Bases, 2. Blenders, 3. Modifiers, 4. Fixers, 5. Florals, 6. Aldehydes. The first two parts provide the background, the second couple provide the character, and the third couple provide the bouquet, of a perfume. The materials used for any one of the different parts are essential oils, synthetic perfumes, artificial perfumes, natural absolutes of flowers, and resins, balms and drugs. The essential oils are the natural volatile essences and terpeneless oils. Synthetic perfumes are the isolated and purified constituents of a natural oil, such as Linalyl acetate which is the principal constituent of Bergamot oil. Artificial perfumes are combinations of synthetic perfumes and the absolute essences of flowers; they faithfully reproduce the natural perfume. A chemist-perfumer may have worked for years on perfecting an artificial perfume. Natural absolutes of flowers are essences obtained from the fresh flower by volatile solvents or other processes. An admirable perfume usually contains the maximum of the formulated allowance for the absolutes. Resins, balms and drugs, when dissolved in the oils and in alcohol, fix the blended perfumes, and prevent one odor from evaporating more rapidly than another. Resins are important in the fixation of perfumes. The famous English perfumer, Mr. Poucher, recommends that fixation be divided into three steps: 1. pre-fixation, deodorization of the alcohol, 2. blending, modifying fixators added when compounding, 3. final fixation, the adding of drugs such as the tinctures of musk, civet, ambergris, castor, and of floral absolutes, to the finished perfume.

Here is a formula for L'Origan, according to the noted French perfumer, Mr R. M. Gattefossé —

253	grams	Base Opoponax	........Bases (containing Blenders and Aldehydes)
160	"	Base Carnation	
245	"	Base Iris	
29	"	Jasmin absolute	........Florals
49	"	Neroli bigarade	
49	"	Bergamot	................Modifier
19	"	Vetiverol	........Fixators
49	"	Tincture Ambre	
49	"	Musk Ketone	
49	"	Tincture Castor	
49	"	Heliotropine	
1000	grams		

To make 1000 cc. of alcoholic perfume: —

10	grams	of the Perfume
990	cc.	of Alcohol 95 per cent.
1000	cc.	

The strength of the perfume may be increased by using: —

25	grams	of the Perfume
975	cc.	of Alcohol 95 per cent.
1000	cc.	

The Bases are compounded according to these formulæ: —

Carnation Base

350	grams	Eugenol
200	"	Isoeugenol
150	"	Acetyl isoeugenol
150	"	Ylang-ylang terpeneless
100	"	Rose absolute
50	"	Santalyl acetate
1000		

Iris Base

830	grams	Methyl ionone delta
10	"	Vetivenol
10	"	Santalyl acetate
100	"	Orris absolute
50	"	Orris concrete
1000		

Opoponax Base

150	grams	Santalol
10	"	Patchouli terpeneless
50	"	Musk Ambrette
150	"	Musk Ketone
100	"	Vetivenol
50	"	Castoreum absolute
100	"	Coumarin
10	"	Rose Bulgarian absolute
360	"	Bergamot
20	"	Neroli Bigarade
1,000		

It is important to mix and dissolve each ingredient added before another ingredient is added to the compound. An amateur perfumer may seldom need to compound a larger quantity of a perfume than 20 grams — enough to make 200 cc. of finished alcoholic perfume. The stock bottle for the compound should be perfectly clean and dry, and accurately counterpoised on the balance; all of the ingredients should be weighed into the bottle, and no oil or solid substance should be permitted to touch the mouth or the inside surface of the neck of the bottle, and no substance should be scattered on the pan of the balance. Such powdery ingredients as Coumarin and Heliotropin and the Musks may be dissolved rapidly, in the compound, by the application of heat; the compounding bottle may be moved over an ordinary, lighted, electric lamp bulb, but the contents of the bottle must not be allowed to get warmer than blood temperature.

Books and periodicals on the subject of perfumery should be interesting to any one wishing to pursue the art. Valuable information can be found in the work of W. A. Poucher, *Perfumes, Cosmetics and Soaps*, in Alfons M. Burger's *Leitfaden der Modernen Parfümerie*, in Otto Gerhardt's *Das Komponieren in der Parfümerie*, and in R. M. Gattefossé's *Formulaire du Chimiste-Parfumeur*. The most interesting periodicals are *The American Perfumer*, the (British) *Perfumery and Essential Oil Review*, and the (French) *La Parfumerie Moderne*, besides Schimmel's annual *Report*.

DITTANY REDIVIVUS

ANNE BURRAGE

IN the 1935 issue of THE HERBARIST, at the end of an article on Origanum Dictamnus L., the author suggested, as a worthwhile project, the introduction to American gardens of this historic herb. It can now be announced with pride that the "sacred healing herb of Crete" has arrived in this country. The following excerpt from a letter from Mrs. Ellery Sedgwick tells the story: —

"When I left for a cruise in the Adriatic last summer I was asked to try and procure seeds of the true Dittany. This little herb is not known in this country and grows only on the hills in Crete. I laid my plans early and spoke to the Greek interpreter on our ship and also to the lady who helps run the cruise. After visiting the excavations and the Museum we arrived at Candia, the port of Knossos, at high noon — and alas, all the shops were closed for the inevitable siesta. Our Greek interpreter failed me in this crisis. Keeping the boatload of impatient people waiting, I bribed a man to see what he could do. Soon he arrived with a bunch of herbs, but I was sure they did not answer the description of the true Dittany. We put off for the ship and on getting aboard, to my joy, Mrs. Anderson presented me with a pot in which grew a little herb which I felt convinced was the Origanum Dictamnus. She had visited a Cretan woman who had been her cook in New York and had now returned to Candia to live. Knowing the " ropes " this woman had procured in one of the provisions stores the only plant of Dittany which they had and which is used on the Island in flavoring drinks.

" The herb was some five inches high and had five or six little purple labiate flowers with protruding stamens at the end of its branches, and resembles exactly the pictures of the specimens of Dittany at the Gray Herbarium of Harvard.

" This little plant travelled on top of a basket from Patmos to Halicarnassus and from Rhodes to Malta, and other ports, and through France on the train. When the flowers fell the bracts became very prominent at the ends of the branches. Nothing could discourage it and it put forth tiny round leaflets when I watered it, which I was told to do every two or three days. The sea voyage and neglect when at home had no effect on its vitality. It seemed to me to have the endurance and persistence of an age-long struggle for existence and its insignificant appearance belied its survival through some three thousand years.

" May it keep these persistent qualities among our richly cultivated and easily discouraged plant life."

. . .

Those of us who love herbs are deeply indebted to our good friend for her contribution of such an historic and legendary plant to our gardens, and it is hoped that, when another season has passed, there may be sufficient cuttings for further propagation.

An additional note of interest has come in a letter from Mrs. G. C. F. Bratenahl of Washington, D. C., who writes: —

Shortly after hearing about this precious plant of the "Righte Dittany," the lost Origanum Dictamnus of Crete (see Mrs. Laurence A. Brown's article, HERBARIST 1935) I came across an article by Eleanour Sinclair Rohde entitled:

"A Collection of Marjorams." (Published in *My Garden*, an English monthly, November 1934 issue.) After speaking of the more familiar varieties of this delightful herb family—those plants that look so "homely and comfortable" in any garden—Miss Rohde writes of Origanum Dictamnus, the "Dittany of Crete" . . . sub-shrubby with downy foliage and pink flowers in August and September. This marjoram, which attains nearly a foot, needs a very warm spot. In Victorian times "Dittany of Crete" was almost as favourite a window plant with cottagers as musk. It is an excellent plant for a cold greenhouse, provided the temperature in winter does not fall below 45 degrees. The compost should be two parts fibrous loam and one part silver sand and leaf mould. During the winter the plants need the minimum of water and as much sunlight as possible. O. Dictamnus can be easily increased by cuttings taken in the spring and struck in sandy soil under a bell glass in the greenhouse. . . ."

Who is familiar with another marjoram that Miss Rohde also includes in her "Collection"? *Origanum microphyllum:* "Marjoram Gentle." It is not listed in Bailey's *Hortus,* Mrs. M. Grieve's *Herbal,* nor does Mrs. Webster mention it. Parkinson describes it with "heads hoary and soft" and Miss Rohde tells us that it has a far sweeter scent than even Sweet Marjoram. "Few aromatic plants have such an exquisite scent. Sweeter and yet also spicier than lavender with just a trace of heliotrope. It needs a light soil and a very sunny sheltered spot." *QUERY:* Could this be, by any happy chance, that "intensely pungent dried marjoram, also from Crete" which accompanied the "lost Dittany" on its long brave journey?

The
HERB FARM SHOP, Ltd.

16, North Audley Street,
London, W. 1., England

Culinary, Medicinal,
Cosmetic Herbs,
"Lady Anne's Preparations"

THE SHOP carries an unexcelled stock of Potpourri, medicinal herbs, creams and lotions, and maintains highly qualified masseuses for herbal Massage.

We are pleased to welcome visitors at **THE FARM** which is just outside the village of Seal, Sevenoaks, Kent.

$1.65 CARRIAGE PAID

BOX OF TEN PACKETS OF DRIED HERBS WITH DIRECTIONS FOR THEIR USE IN COOKING

CARRY-ON SHOP
65 CHARLES STREET
BOSTON, MASS.

WE USE AND SELL
our *home grown, home prepared* HERBS
at our two Tea Rooms

THE HERB FARM
Hacklebarney Road, Chester, New Jersey
and

THE HERB KETTLE
Acacia Road, West Palm Beach, Florida

where we specialize in *home-cooking* delectably flavored with our delicious *blends* of cut and *powdered herbs* which once used become indispensable in your kitchen

Write for our booklet describing our:

Spiced Herb Powder
Old Fashioned Herb Powder
Tarragon Herb Powder
Poultry Stuffing Herbs
Salad Herbs
Omelette Herbs
Herb Bouquet
Herb Vinegar
Tomato Cookery Herbs
Fish Cookery Herbs

FINE PRINTING

Estimates will be gladly furnished for all Garden Club needs

Programs

Stationery

Labels

Year Books

THE PERKINS PRESS

Topsfield Massachusetts

Old English Herbs
from
the Olitory at Stonehurst,
Shanty Bay, Ontario, Canada.

Balm	Nasturtiums (Pickled)
Basil	Parsley
Dill (Seeds)	Summer Savory
Fennel	Tarragon (Leaves
Marjoram	" (Ground)
Mint	Thyme

Sage

The Herbs furnished from the Olitory at Stonehurst, may be used dry at the table to flavour soups, sprinkle potatoes, etc., the same as you use pepper and salt.

These herbs are grown in suitable soil, in sunshine or shade as is most fitting, harvested at the right time, dried in the proper way, carefully winnowed and finely powdered. They are absolutely pure and much stronger flavoured than herbs usually obtainable.

---

1 Jar.................... .50
Box of 6 jars........... $2.50

These herbs make an attractive and useful Christmas present. A "Herbal" of historical value and containing many receipts from the old herbals, beginning with "Dioscoridies" of the first century A. D. up to the XVIII century, is sent with each box.

Toronto Office

THE GREENMAN

238 Bay Street Elgin 4180

LINNÆUS
1774

MORE PLANTS
for the
HERB GARDEN

Those who saw our display of Herbs at the Massachusetts Horticultural Society on August 17, 1935, for which we were awarded the Society's Gold Medal, are aware of the unusual variety of Herbs we are growing.

You will want a copy of our booklet "Plants for the Herb Garden" which includes over seventy-five varieties such as Tarragon, Applemint, Rosemary and a host of other plants practical for their flavoring, seasoning, or medicinal properties or equally valuable for the fragrance or sentiment they bear.

Many like Germander, Winter Savory, and Hyssop add their neat foliage to the group particularly in trim "Knot Gardens" and perennial borders.

If limited for space, use them as a part of your perennial border. Many varieties are richly fragrant, while others have showy flowers which can be used to advantage in such an arrangement.

Let us remind you, too, that we grow Evergreens, Trees, Flowering Shrubs, Vines, Roses, and Perennials in a wide variety, enabling us to meet all your garden requirements. These are illustrated and described in our 96 page Handbook for 1936 — also mailed on request.

BAY STATE NURSERIES

Incorporated

NORTH ABINGTON MASS. and FRAMINGHAM MASS.

Telephone Rockland 26 — Framingham 6191

The HERBARIST

A Publication of

The Herb Society of America

No. 3

For Use and for Delight

BOSTON, MASSACHUSETTS

1937

The Herbarist for 1937
Written and Published by
The Herb Society of America
Price $1.00

A few remaining copies of the 1935 and 1936 Herbarists may be purchased at $1.00 each.

Address

The Herb Society of America
Horticultural Hall
300 Massachusetts Avenue
Boston, Massachusetts

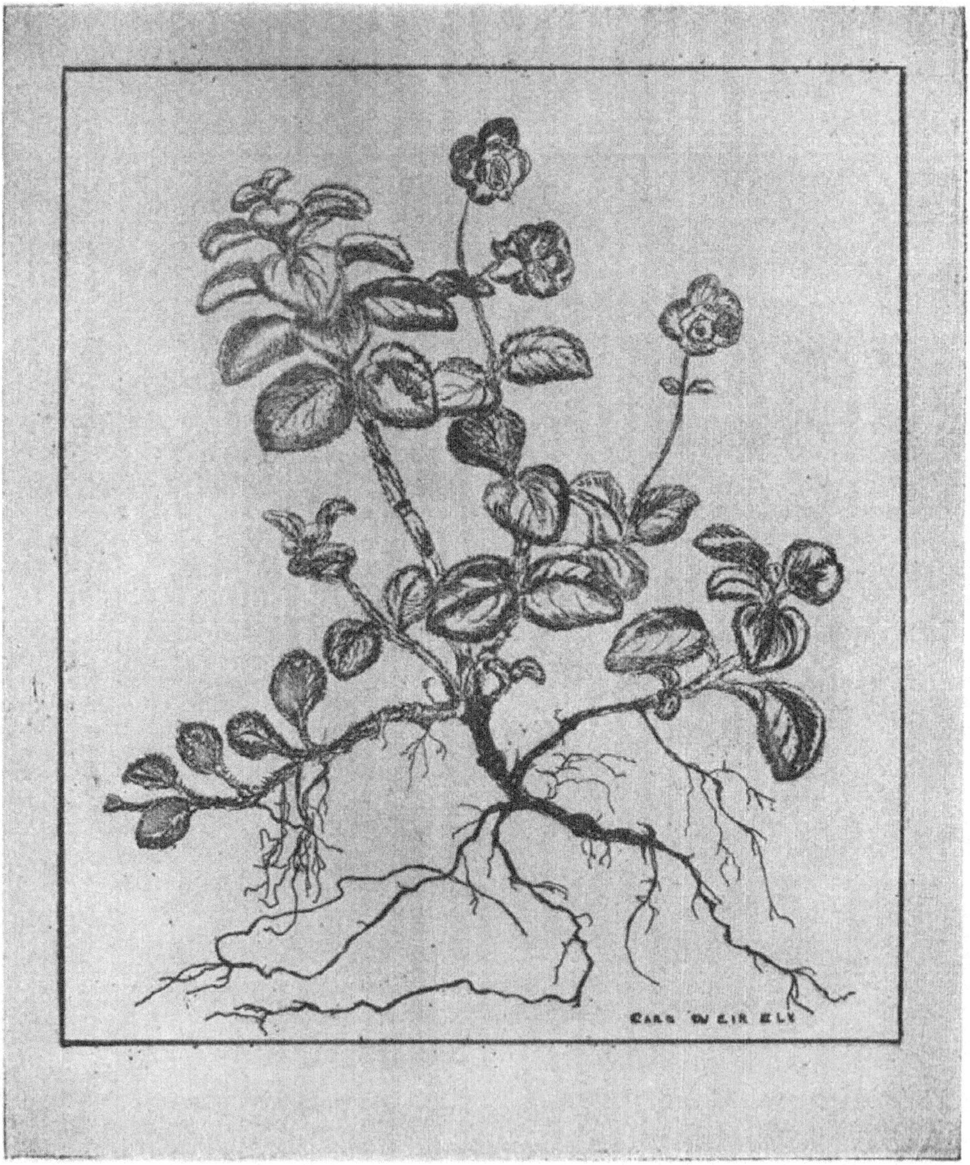

ORIGANUM DICTAMNUS L

An original etching by Caroline Weir Ely

This plant was propagated in Wenham, Massachusetts, from the root brought from Crete in 1936.

CONTENTS

Ephedra Sinica, one of the chief sources of ephedrine

EPHEDRA

Its History, Distribution, Constituents, Uses and Cultivation

HEBER W. YOUNGKEN, PH.D.
PROFESSOR OF MATERIA MEDICA, MASSACHUSETTS COLLEGE OF PHARMACY

THE records in the Pên ts'ao or Chinese herbals show that Ephedra has been used as a medicine in China for more than five thousand years. Its curative properties are still highly regarded by the Chinese, while these have been only comparatively recently known to the people of other countries owing to the language barrier. At the present day it is sold in Chinese drug and herb shops, both in the form of green to greenish-brown sticks which represent the dried stems of the plants, and in the form of dried roots. The stem drug is known to the Chinese as "Ma huang" and the root drug as "Ma huang kan." The stem of a plant found to yield ephedrine was discussed by Li Shih-chen in his " *Pen Tsao Kang Mu*" of 1596 A.D. (1).

The genus *Ephedra* is represented by at least forty valid species of straggling shrubs which inhabit the warm, arid regions of six continents, namely, Asia, Africa, Europe, North America, South America and Australia.

The Ephedras are mostly low-branched shrubs of prostrate or climbing habit from whose woody root systems emanate slender, jointed, green, aerial stems with opposite or fascicled, slender, jointed, green branches, the entire aerial growth resembling that of the horsetails. The green stems and branches bear minute, scale-like, simple, exstipulate leaves at their nodes which are usually two- or three-parted, more or less connate, and sheathing. The flowers are unisexual and mostly diœcious and occur in small, usually axillary aments with decussate bracts. The staminate flowers possess two to eight stamens which are solitary or united into a column and surrounded by a membranous perianth. The pistillate flowers consist of an erect naked ovule seated within an urn-shaped perianth. The fruiting cones contain one, two or three pistillate flowers. The fruit is a red, berry-like syncarp containing a single, bi-convex, elliptic or ovate, albuminous seed.

According to O. Stapf (2) Ma huang in the old Chinese literature refers to more than one species of *Ephedra.* This is borne out by the statement in the *Pen Tsao Kang Mu* by the wide distribution therein cited.

The orientals employed Ma huang as a diaphoretic and circulatory stimulant as well as in their treatment of asthma, bronchitis and urinary diseases. They use the roots as an antidiaphoretic in profuse sweating.

The following species of Chinese Ephedra are cited in the literature:

1. *Ephedra equisetina* Bunge, known as "Mu Pen Ma huang," "Ma Wei Tsung," "Sung," "Mu Tsei Mahuang," and "Mahuang." It grows in clay or sandy soils chiefly in northern Hopei from the seacoast at Shankaikwan to Kalgan, west to Paotowchen, north to Shensi and Shansi.

2. *Ephedra sinica* Stapf, known as "Tsaopen Mahuang." This grows in sandy or clay soil from the coast at Shankaikwan to Kalgan, principally in northern Hopei. It is also found in Pao t'on and in the mountains of Shansi and Shensi and between the northern parts of the boundaries of Shansi and Hopei.

3. *Ephedra distachya* Linné (*E. vulgaris* Rich.) claimed by Read (3)

and other authors to be the same as *Ephedra vulgaris* var. *Helvetica* Hooker and Thompson.

4. *Ephedra Gerardiana* Wall., a native to western China and Sikkim in the eastern Himalayas of India.

5. *Ephedra intermedia* Schrenk. et Meyer (*E. pachyclada* Boiss) found in scattered areas of China, India and Central Asia.

6. *Ephedra monosperma* Meyer, found in China, Mongolia and northern Asia.

7. *Ephedra minuta* n. sp. R. Florei, a low growing plant with solitary flowers and short internodes, native to China.

8. *Ephedra lomatolepis* Schrenk., native to the Sungarian steppe.

European Species and Varieties of Ephedra

1. *Ephedra helvetica,* C. A. Meyer, occurring in Italy, France and Switzerland. It yields pseudoephedrine, but no ephedrine.

2. *Ephreda distachya* Linné, which also occurs in China.

3. *Ephedra Villarski,* a variety of *E. nebrodensis* Tineo, occurs in Italy.

4. *Ephedra altissima* Desf., a species found in Switzerland.

5. *Ephedra Andina,* a variety of *Ephedra americana* Hanb., occurring in Switzerland.

6. *Ephedra nebrodensis* Tineo occurs from the Mediterranean region to the Himalayas in Asia.

African Species of Ephedra

1. *Ephedra altissima* Desf.

2. *Ephedra alata* Decne.

North American Species of Ephedra

Several species of *Ephedra* grow wild in North America. Most of these are found in the southwestern portion of the United States and in Mexico. The North American Indians made use of a number of these and the white settlers learned the medicinal uses of the plant from them. American species abound in tannin content and it is upon this constituent that their medicinal value depends largely. According to Loew (4) extractions of American species of *Ephedra* made with water show the presence of tannin, pectin and tartaric acid. The tannin is glucosidal and upon treatment with an acid yields sugar and a reddish powder.

The Indians and white settlers prepared a tea from the green stems of various species which they used for venereal diseases and other ailments.

Nine species of *Ephedra* grow in the United States, all being found in the western and southwestern part of the country. Their distribution in general is as follows:

1. *Ephedra antisyphilitica* Berland, which occurs in New Mexico, Arizona and Southern California.

2. *Ephedra aspera* Engelm., found in California, Pyramid Lake, Nevada, in Colorado and near El Paso, Texas.

3. *Ephedra californica* Wats., occurring in Colorado and California, especially at Point Loma, San Diego.

4. *Ephedra trifurca* Torr., found in Colorado, Arizona and California.

5. *Ephedra viridis* Coville, occurring in Nevada, Arizona, Utah and southern California.

6. *Ephedra nevadensis* Wats., found in Arizona, New Mexico, Colorado, Utah, Nevada and California.

7. *Ephedra pedunculata* Engelm., occurring in western Texas.

8. *Ephedra torreyana* Wats., occurring from New Mexico to southern Utah.

9. *Ephedra peninsularis* I. M. Johnston, found at Baja, California, and thence northward to Calmalle and Cedros Island.

C. Nielsen, H. McCausland and H. C. Spruth assayed specimens of American species, employing the U. S. P. IX. method for Belladonna Root but did not find any alkaloid present (5).

The Indian tribes of the Southwest were familiar with several of the native *Ephedra* species. The Coahuilla Indians roasted the seeds of *Ephedra nevadensis* as food and a decoction of the green stems was used as a cooling drink.

The Panamint Indians, according to F. V. Coville (6), roasted and ground the seeds of this species, making them into a bitter bread.

D. P. Barrows (7) gives the following account of the use of Ephedras by the Coahuilla Indians of southern California: "Besides these cooling drinks the Coahuillas have a tea plant that is very much prized. This is the *Ephedra nevadensis* Wats., called tu-tut. The Ephedra has, I believe, well-known medicinal properties; certainly the tu-tut is much valued by the Coahuillas. Branches of these twigs are almost universally tucked among the thatchings of every jackal or packed away in baskets or 'olla.' The tea steeped from it forms a pleasant and refreshing beverage."

The Arizona Indians are said to have used species native to that state as a diuretic. The Zuñi Indians, according to M. C. Stevenson (8), made a tea from the plant minus the root which was drunk during the first stage of syphilis.

The following species of Ephedra are reported in the literature as growing in Mexico:

Ephedra antisyphilitica Meyer, occurring in Coahuilla and Durango.

Ephedra aspera Engelm., reported in Coahuilla.

Ephedra californica Wats., in Lower California.

Ephedra pedunculata Engelm., occurring in Chihuahua, San Luis Potosi and Coahuilla.

According to P. C. Standley (9), *Ephedra aspera* and *Ephedra pedunculata* are sold in the markets in Mexico as remedies for pneumonia.

A. Schneider (10) reported that decoctions of *Ephedra californica* Wats., were used by the Indians and Spaniards as a tonic and blood purifier.

The Constituents of the Ephedras

The constituents of the American species are discussed in an earlier part of this paper.

The active principle of most of the Asiatic species of *Ephedra* has been found to be the lævorotatory alkaloid ephedrine. It was first

Ephedra Gerardiana, a source of ephedrine

isolated by Nagai and his associates from material Nagai called *Ephedra vulgaris* Rich. var. *helvetica* Hook. and Thompson (11), a plant which has more recently been determined as *Ephedra distachya* Linné. According to E. R. Miller (12) the European variety of this species (from Switzerland) yielded only pseudoephedrine. Chen in 1923 again isolated ephedrine from *Ephedra* plants purchased in Peking and its vicinity. In 1926 K. K. Chen and C. H. Kao (13), stated—"it appears, therefore, probable the plant *Ephedra vulgaris* var. *helvetica* yields ephedrine when grown in China, but pseudoephedrine when grown in Europe." However, B. E. Read and Chih-Tung Feng (14) show that ephedrine is found both in the Swiss and Chinese plants. Ephedrine is closely related chemically to epinephrine or adrenalin found in the medulla of the adrenal gland and also to tyramine found in Ergot.

The same authors in a subsequent paper have compared the seasonal content of ephedrine in *Ephedra sinica* and *Ephedra equisetina* monthly from mid-May to mid-October and have found a decided increase in the amount present as the plants mature. In the case of *E. sinica* the percentage of total alkaloid of plants gathered from May 14-16 was 0.438, of which 75-85% consisted of ephedrine and 15-25% pseudoephedrine, whereas plants gathered from October 1-10 yielded 1.315% of total alkaloids, of which 80-85% represented ephedrine and 15-20% represented pseudoephedrine. In the case of *Ephedra equisetina* plants gathered from May 14-16 the percentage of total alkaloids was 0.614, of which 85-90% represented ephedrine and 10-15% pseudoephedrine, whereas plants gathered from October 1-10 yielded 1.754% of total alkaloids, of which 85-90% represented ephedrine and 10-15% represented pseudoephedrine (15). It is obvious from these seasonal tests that autumn is the best time for the collection of Ephedra in China.

Modern Medicinal Uses of Chinese Ephedra and Preparations

In earlier parts of the paper the writer has discussed uses of various species of Ephedra in domestic medicine. In rational medicine today use is made of the crude drug, the alkaloid ephedrine, its salts ephedrine sulfate and ephedrine hydrochloride, and of various galenical preparations of these. The alkaloid and its salts became official in the last edition of the *United States Pharmacopoeia,* while a number of preparations including ampuls of ephedrine hydrochloride, ampuls of ephedrine sulfate, syrup of ephedrine sulfate, solution of ephedrine sulfate, tablets of ephedrine hydrochloride, as well as two nasal inhalants or sprays containing the alkaloid, namely, ephedrine spray and compound spray of ephedrine, became official in the latest edition of the *National Formulary.*

As first shown by Chen and his associates, ephedrine is a stimulant to the sympathetic nervous system, acting somewhat like epinephrine but exhibiting less intense though more prolonged effects than those of epinephrine. It possesses three advantages over epinephrine, namely, it can be given by mouth, its action is more prolonged, and it is safer in that its hypersusceptibility reactions are fewer. Ephedrine is used locally to shrink mucous membranes in the treatment of hay fever, rhinitis and small nasal hemorrhages in the form of a solution or as a spray, jelly or ointment, and as a mydriatic for producing dilatation of the pupil in eye diseases in the form of the solution. It is administered internally in certain cases of surgical shock and circulatory disturbances, in bronchial asthma, in cocainism, etc., and for controlling the severity of the paroxysms in whooping cough. The official syrup of ephedrine sulfate as well as a syrup made from the crude drug are on the market as remedies for coughs.

Cultivation of Ephedra

While a few species of *Ephedra* have been propagated on a very small scale in scattered parts of this country mainly in hot houses, it remained for Christensen and Hiner, in 1936, to prove that Chinese *Ephedra sinica* can be successfully grown as a domestic drug crop in this country (16). Seed for trial planting was obtained from the Peking Union Medical College in 1929 and the seedlings obtained therefrom were transplanted in pots in May 1930 and later transplanted into the medicinal plant garden at Brookings, S. D., in June 1930. The first stem cuttings were made during the fall of the following year and the plants have become established there, yielding yearly crops. Some of the plants have been transplanted to a test plot in the Bad Lands section of South Dakota. Here the stems and roots were found to be hardy and perennial. The plants in both plots rooted deeply and produced viable seeds.

The aerial stems should be harvested in late September or just before too heavy frosts may be expected. The crop is mowed down like hay, allowed to cure in the field and then baled for shipment.

While the highest yield of the South Dakota grown drug was 0.366% of ephedrine, which is lower than that reported for that of Chinese grown *E. sinica*, nevertheless, the undertaking must be regarded as successful, for it proves that ephedrine containing Ephedra can be successfully grown in the United States, and that we will not have to be altogether dependent upon supplies of the foreign drug in periods of emergency.

Bibliography

1. *Pen T'sao Kang Mu, Sect.* 15, *monog.* 20, 1597.
2. *Kew Bull.*, 1927, 133.
3. *B. E. Read.* The Botany of Mahuang. Part 2, 16, 1930.
4. O. *Loew. Investigation upon mineralogical and agricultural conditions observed in portions of Colorado, New Mexico and Arizona in* 1873, *Wheeler's Rept.* 7, 611-12, 1875.
5. *Jour. Am. Pharm. Assoc.*, 16, 288-94, 1927.
6. *Am. Anthrop.* 5, 1-353, 1892.
7. *The ethnobotany of the Coahuilla Indians of S. California*, 73-74, 1900.
8. *Ethnobotany of the Zuñi Indians*, 13 *Ann. Rept. Bur. Am. Ethnol.*, 49, 1915.
9. *Contrib. U. S. Natl. Herb.*, 23, 63-64, 1920.
10. *Calif. State Board Forestry Bull.* 2, 95-96, 1912.
11. *Pharm. Zeitung* 32, 700, 1887.
12. *Arch. Pharm.* 240, 481-83, 1902.
13. *Jour. Am. Pharm. Assoc.* 15, 625-39, 1926.
14. *Chinese Jour. Physiology*, 1, 297-304, 1927.
15. *Ibid.* 2, 87-96, 1928.
16. *Jour. Am. Pharm. Assoc.*, 25, 969-73, 1936.

PREHISTORIC MAN'S INTEREST IN PLANTS

G. B. MAC CURDY

MAN'S interest in plants is as old as the human race. For hundreds of thousands of years man was absorbed in the quest for food. He learned by experience which foods were best suited to his needs and the knowledge thus gained was passed on by word of mouth from generation to generation and from age to age. In such circumstances it is easy to see how myth, legend and tradition might grow up relative to the virtue of certain plants.

Our knowledge of man's use of plants in prehistoric times is limited because plant remains are more perishable than the skeletal remains of animals on which he fed. Ideal conditions for the preservation of plant remains are found in peat bogs and lacustrine deposits. Wherever these deposits are found to contain evidences of human occupation, such as pile villages, one may expect to find remains of plants used by the peoples of that time.

The first scientific exploration of pile-village sites was begun nearly one hundred years ago. The oldest of these sites belonged to the Neolithic Period dating from about 5000 B.C. Besides implements used in hunting and fishing, there were found pottery, milling stones, spindle whorls, woven fabrics and even loaves of a coarse kind of

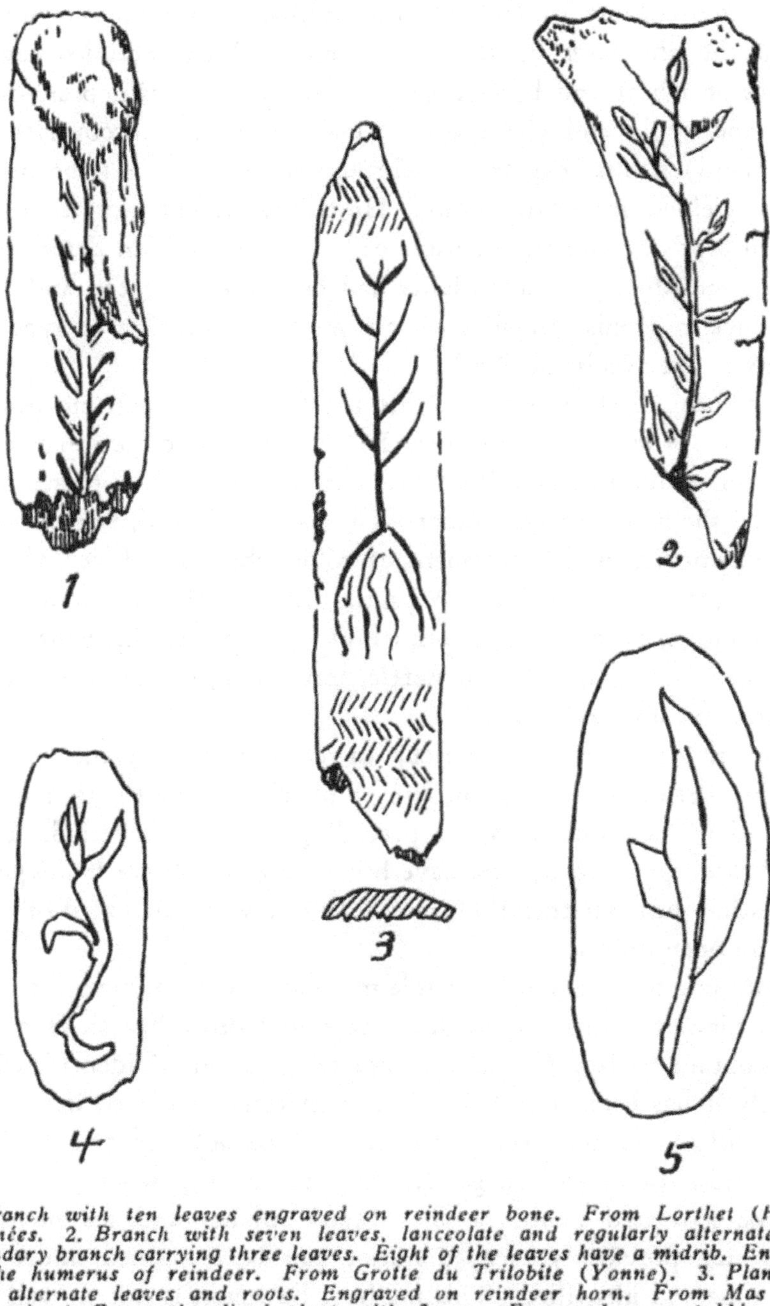

1. *Branch with ten leaves engraved on reindeer bone. From Lorthet (Hautes-Pyrénées. 2. Branch with seven leaves, lanceolate and regularly alternate, with secondary branch carrying three leaves. Eight of the leaves have a midrib. Engraved on the humerus of reindeer. From Grotte du Trilobite (Yonne). 3. Plant with eight alternate leaves and roots. Engraved on reindeer horn. From Mas d'Azil (Ariege). 4. Conventionalized plant with flower. Engraved on a pebble. From Gourdan (Haute-Garonne). 5. Leaf engraved on pebble. From Gourdan (Haute-Garonne).*

bread made without leaven. Bones of animals and plant remains indicated the character of the food supply which included several varieties of wheat and barley, spelt, millet, oats, lentils, peas, broad beans, the apple and the grape. Since these first discoveries (in Switzerland) similar exploration throughout Europe and the British Isles has yielded important results. Up to the present more than two hundred kinds of plants, exclusive of cereals, have been determined. At least one-third of these plants had been domesticated and their occurrence in prehistoric sites has an important bearing on race migrations of the Neolithic Period.

In addition to their use of plants for food, the Neolithic peoples used plants for medicinal purposes. It is interesting to note that many wild species found in Neolithic sites can be obtained in present-day chemists' shops on the Continent. They are employed, for the most part, as remedies in the treatment of colds, chills and fever, though some are used as astringents. A partial list includes: agrimony, angelica, boneset, burnet, cinquefoil, fumitory, hemlock, henbane, horehound, mint, mullein, mustard, nettle, opium poppy, sage, strawberry, valerian, vervaine and viola.

From the number and variety of domesticated plants of the Neolithic Period, one is inclined to think that domestication might have had its beginnings in the preceding Mesolithic Period. Flint sickle blades, polished by use, have been found in Mesolithic deposits in Palestine, but whether the grain they cut was cultivated or wild is still an open question.

Until some ten years ago our sole method of determining the presence of plant remains was to sieve material from a hearth or refuse pile to obtain seeds sufficiently complete to admit of identification. Recently it has been found that if this material is soaked in a weak solution of nitric acid (or potassium hydroxide) for twenty-four hours, seeds, leaves and twigs will float to the top where they can easily be separated from the silt. Many smaller and fragmentary seeds which previously escaped notice can be examined under the microscope and identified. This method, however, is not so well developed as that of pollen analysis, by which we will be enabled to reconstruct the flora of even more remote periods of prehistory. An example of this is the recent identification of twelve kinds of plant remains, among them alder, from a lacustrine brick earth. These remains were associated with stone tools made by an old paleolithic race (late Acheulian or early Mousterian) which lived during the last interglacial epoch about 100,000 years ago.

There is still another method of approach to the problem of prehistoric man's interest in plants, *viz.*, an examination of plant representations in his art. The oldest known records of this kind were left by the Magdalenians, the last of the paleolithic races who preceded the Mesolithic peoples previously mentioned. These Magdalenians lived during the closing phase of the last glacial epoch, some 20,000 years ago. From fossil remains it is evident that a true northern tundra flora existed, such as pine, polar willow, dwarf birch, mountain avens, knotweed, etc. There were long, cold winters and short summers, but wherever the tundra was free of ice it was carpeted with grass and shrubs and along the glacial-fed streams grew sparse and stunted forests. The fauna, dependent upon flora and climate, was arctic, including reindeer, mammoth, musk ox, bison, cave bear, a species of horse, etc. These animals, in addition to edible plants, comprised the Magdalenian sources of food. Their quest for food is reflected in Magdalenian art. Animals were not easily caught by a hunter limited to primitive means, therefore he invoked the aid of magic. Many of the animal figures engraved, painted, or in the round may be construed as prayers for success in the chase or to assure continuation of the food supply through increase in the herd.

A careful examination of all available records shows that plants also were reproduced in cave art—a fact we have been too prone to overlook. Some fifty representations are known; all are from the Magdalenian stations of southwestern France. They comprise engravings, figures carved in relief, in the round, and paintings. The subjects are grass through which animals are moving, trees or shrubs, branches, leaves, flowers and conventional designs derived from flowers. It would be difficult to identify the species which the artists represented. It is significant that the authors of these plant figures were the first to make representations which have been accepted as sun symbols.

WEATHERED OAK HERB FARM, INC.
BRADLEY HILLS
BETHESDA, MARYLAND

THE WEATHERED OAK HERB FARM

FLORENCE BRATENAHL[1]

AS you approach the farm up the long hill to its entrance path, you see the little house in profile. There is the massive chimney, of worn stone, with the crude tool-marks of two centuries ago: the whole strength of this broad shaft not only gripping the earth but growing out of the ground as though it had always been here. And yet the structure as you see it today: chimney, gable ends, tiny dormers, long sweeping roofline to the west towards the planting fields just beyond; the old farm bell, wooden plough, worn doorstep, did not exist here a few months ago. Not a stone. Not a timber. This may be difficult to believe when you notice the construction of solid oak, the tone and texture of ancient hand-hewn beams and braces, the sag as of burdened age in the long slant of that shingled roof, not to mention its moss and its lichen!

The gathering together of all this interesting and worthwhile material had been an adventure in itself. A strong wind last spring hurrying across this peaceful Maryland countryside seemed to blow down just the kind of things we were looking for: an old hayrick, one or two old barns and a corncrib; while a little human breath, "a quiet breathing," was all that was necessary to blow down and then to collect more timbers, more old roofing, rafters and sheathing—but it must be *oak*.

It was one thing to collect and assemble this oddly assorted material. Quite another thing to rebuild with it. Most of it was amazing in its sheer solidity and strength in spite of so many generations of

[1] Mrs. Bratenahl is Vice-President of the Herb Society of America, one of its founders, and the designer of its seal. This article is an excerpt from the complete story of Weathered Oak Herb Farm appearing in the current number of the *Bulletin of The Garden Club of America.*

usage. In fact it was so sound, so hard, so difficult to handle, hammer and saw, that no first-class carpenters could be found who could be persuaded to undertake the work. So the actual building was executed by farm labor under the constant supervision of those who were responsible for the designs. Of necessity the work progressed slowly but at last, after three months of effort, the little house, built on its two different levels, was done. One large room, several smaller ones, an outside stairway leading to a drying loft above, two chimneys: the large stone one and, at the opposite end of the house, a small brick one for furnace purposes. Even an old-fashioned cellar door—but too small to slide down! Everything complete except an adjacent planting of intimate little herb gardens which the spring would help to create. We could now look about us and re-study the hillside as a whole, including those wide acres of farmland, deeply ploughed, most favorable for healthy growth.

Indoors we hung soft brown linsey-woolsey curtains at small casement windows; and, on the many oak shelves, placed jars of carefully prepared dried herbs, our own blends, from a first harvest here. Then there were all those jams, honeys and kindred products. And it was fun to carry armfuls of garden books, old ones, ancient herbals, crisp new ones fresh from the publisher. Just to handle the books, a delight in the tiny Book Room, with its mellow pine furnishings, cupboards, shelving, low bench, bowl of apples, and its single small chair. Beside it is a small wooden chest with paint faded, lock broken. It must have once held a child's secrets and treasures. Well, they are ours now and the childlike joy of so many years ago is the best part of this room's heritage.

It was December when we lit the first fire on the hearth and started the kettle boiling. Great logs contributed to the blaze and to the kettle's low murmur—surely the most ancient folk-song of any land. Presently, with heat extending to the far corners of the room, slumbering honey-bees in the very heart of this house of oak awakened. They seemed a happy omen. Instinctively the little place caught up their song with its old traditions. From that moment the Herb House began to serve its purpose as a natural center of activities of a newly developed farm for the growing of herbs.

ANGELICA ARCHANGELICA THE GREAT UMBELLIFER

EDGAR ANDERSON

OF all the angelicas, and there are around fifty of them in the northern hemisphere, only two are likely subjects for the American herb-garden: *Angelica Archangelica,* the cultivated Angelica and *Angelica atropurpurea,* the wild American species sometimes known as masterwort. They are lusty fellows, given the rich moist soil they love; spending a youthful year or so to develop a rosette of leaves and a parsnip-like root and then in a single season dramatically bursting up six to ten feet high to flower and set seed and die. Their leaves, rather like a cruder celery leaf, rise from bulbous bases so wide that they encircle the hollow stem. The flowers are small, but they are bunched in tight little pompons and groups of these pompons are compounded into a large sunburst. This whole inflorescence is technically known as an umbel of umbels, or a compound umbel, a dull enough name for such a fascinating structure, which in its main lines resembles a bursting sky-rocket.

In flower, in leaf, and in general habit these two species of angelica are very much alike and rather different from most of the other species of the genus. For one thing they are smooth-stemmed and smooth-leaved, as smooth as celery, while many of the other species,

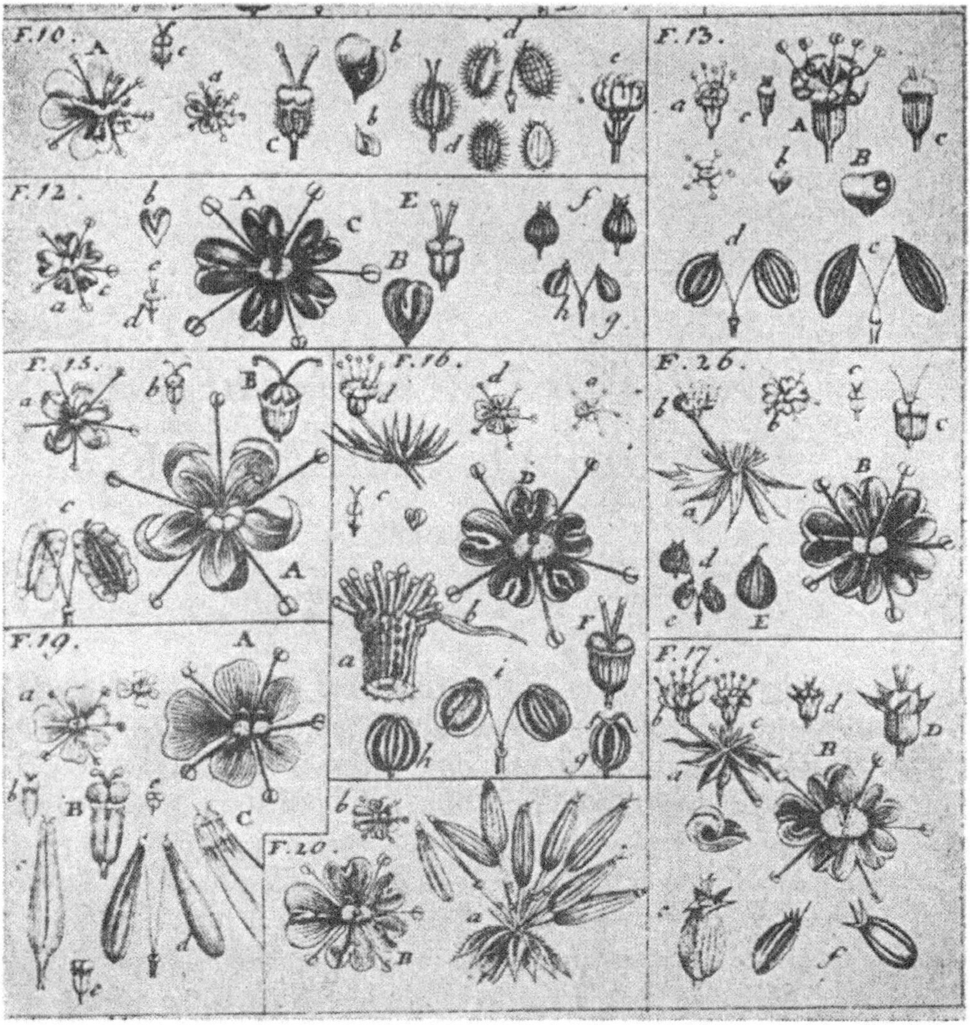

Flowers and Seeds of various Umbelliferae
(*All very much enlarged*)

From Hoffmann, *Genera Umbelliferarum*, 1816, vol. 1, Tab. 1A.

Fig. 10. *Daucus Carota* (carrot). *a.* ray florets; *A.* disc florets; *b.* petals; *c, C.* ovary; *d.* fruit and seed; *e.* center floret.

Fig. 12. *Pimpinella Saxifraga* (burnet-saxifrage). *a, A.* florets; *b, B, c, C.* petals; *d, E.* ovaries; *f.* fruit; *g.* seed.

Fig. 13. *Anethum graveolens* (dill). *a, A.* florets; *b, B.* petals; *c, C.* ovaries; *d.* seeds; *e.* seed of fennel for comparison.

Fig. 15. *Angelica sylvestris* (European wild angelica) *a, A.* florets; *b, B.* ovaries; *c.* seeds.

Fig. 16. *Selinum Carvifolia* (milk-parsley). *a, b,* stem and involucre; *c.* petals and ovary; *d.* floret and involucel; *D.* floret enlarged; *e.* center floret; *F.* ovary; *g, h.* fruit; *i.* seeds.

Fig. 19. *Anthriscus sylvestris* (chervil) *a, A.* florets; *b, B.* ovaries; *c, C.* fruit; *d.* seeds; *e.* sterile ovary.

Fig. 20. *Chaerophyllum temulum* (chervil) *a.* umbel with seeds; *b, B.* florets.

Fig. 26. *Sium angustifolium* (water-parsnip) *a.* involucel; *b, B.* florets; *c, C.* ovaries; *d.* fruit; *e, E.* seeds.

as for instance the wild European angelica, *Angelica sylvestris,* are more or less hairy. The garden angelica, *A. Archangelica,* differs from its American cousin, *A. atropurpurea,* in being a little less coarse, much less highly colored and in having a more delicate aroma. The stems and leaf bases of *Angelica atropurpurea* are strongly flushed with dark red-purple; those of *Angelica Archangelica* are green or at most streaked with brownish red. The American species has a rank, medicinal odor, the European a delicate and attractive one, strongly reminiscent of Benedictine and similar liqueurs (a natural enough association, since the seeds of angelica are one of the ingredients of many such *apéritifs*).

Angelica is interesting as being one of the relatively few garden plants which actually originated in north Europe. The species is native there from the Faroes to Lapland and since ancient times has had a popular reputation as a cure-all. Its very names, "Heilige Geist," "Angelwurz," and the name "Angelica" itself, bear tribute to the veneration in which it has been held. To the modern herb gardener it is of interest chiefly by reason of this history and for the succulent stems which are usually candied.

Angelica belongs to one of the most distinctive of the great plant families, the Umbelliferæ ("The Umbel Bearers") as do also dill, lovage, chervil, parsley and other lesser herbs. Not only in their umbel of umbels, but in many another character do these various herbs show their relationship. The leaves all possess the same pattern, being merely more or less cut as the case may be, finely cut in parsley, superfine in dill, and coarse in celery and lovage. Through the whole family there runs a very similar set of odors and flavors; celery to carrot, carrot to parsley, parsley to dill, dill to fennel, and fennel to sweet cicely. The flowers are similar throughout, not only in their arrangement but in the pattern of the actual florets themselves. Yellowish, greenish, or white, they have five tiny petals and five spreading anthers and vary only slightly from species to species and from genus to genus. On the accompanying plate, for instance, see how closely similar are the florets of carrot (*fig.* 10), chervil (*fig.* 20), and water-parsnip (*fig.* 26). It is not until they go to seed that the Umbellifers reveal the differences of opinion that really do exist within the family. Then we find the carrots distinguished (again we refer to the plate) by their rough, barbed fruits, the chervils by long pendants and the water-parsnips by chunky little pills.

The fruits of the angelicas are soft, corky and gray-brown, shaped something like a pair of lady-bugs but two or three sizes larger. Around the sides (*fig.* 15c) is a flat margin, perfectly plain in the wild American angelica, slightly waved in the garden angelica, and still more undulate in the wild *Angelica sylvestris* of European hedgerows which is illustrated in the plate. From front to back the fruit has three corky ridges. These are slightly conspicuous in *Angelica sylvestris*, decidedly winged in *A. Archangelica*, and are even more prominent in *A. atropurpurea*, whose fruits are also somewhat larger, as we might anticipate from so coarse a plant. So much alike are these two latter species that by some botanists they are split off into a genus by themselves. We may therefore find *Angelica Archangelica* L. hiding under the name of *Archangelica officinalis* Hoffm., while our own *Angelica atropurpurea* L. becomes *Archangelica atropurpurea* (L.) Hoffm. But these are minor matters and whether we join Linnæus in calling them angels, or with Hoffmann raise them to the archangels, they remain among our most imposing herbs and are the most magnificent of the cultivated Umbellifers.

BURNET — A PENCIL DRAWING BY LOUISE MANSFIELD
Owned by The Herb Society of America.
One of the plates to be included in "An Artist's Herbal" (Macmillan, 1937).

SOME SAXON HERBALS

ELEANOUR SINCLAIR ROHDE

AMONGST the countless, priceless MSS. in the British Museum, those that have always fascinated me most are the Saxon Herbals. Unlike the gorgeous illuminated treasures, these humbler MSS. are for the greater part devoid of illustrations and again, unlike the former, their very appearance bears silent testimony to the fact that, so far from being carefully guarded and seldom handled, these early herbals were in constant use and much thumbed. Quite apart from their intrinsic interest, this gives them, to my thinking, a human interest. When I turn over the pages, I like to think of the Saxon leeches who handled these documents and consulted them, possibly daily, and to imagine what manner of folk have owned these volumes through the centuries.

When I turn the pages of the *Leech Book of Bald*, it thrills me to think that it is at least possible that King Alfred may have handled this very same book. The writer was evidently so close a friend of the King that he had access to Alfred's correspondence, for he quotes various prescriptions sent to the King by Helias, Patriarch of Jerusalem. It is a beautiful volume and centuries later it was owned by the monks of Glastonbury. Centuries of wear have not impaired the stout vellum on which it is written and the writing is a magnificent specimen of Saxon penmanship—

a big, bold writing. And fortunately the name of the writer has been preserved for us. There is a verse in Latin, at the end of the first part, stating that he was one Cild, who had been employed by Bald (presumably the leech) to write. There are curious marks on a great many of the pages, but no one can solve what they mean.

In the MS. there are references to two other leeches, Dun and Oxa, who had given Bald prescriptions. The leech's position in those days was not an enviable one for, unlike modern doctors, who are regarded as final authorities by at least the majority of their patients, Saxon doctors were subjected to what can only be described as the unfair competition of the higher clergy, who were credited with miraculous powers of curing all sorts of diseases. Not surprisingly, therefore, the leeches turned to the old heathen superstitions, knowing full well that in spite of nominal Christianity the people still feared their old gods. It is this fact that gives this volume its peculiar interest, for Bald incorporated in it ancient herb lore handed down from what dim past centuries it is possible only to imagine. In one incantation the names of ancient heathen gods—Tiecon and Leleloth—are associated with the names of the Four Evangelists. These pages carry us back still further to the old beliefs in malevolent unseen beings always at warfare with mankind, and there are still remote parts of these islands where these beliefs are far from extinct. Victims of disease were regarded as elf shot, and on herbs, administered with incantations, our remote ancestors relied to undo their baleful effects. Some of these incantations are suggestive of nursery rhymes. Nor are beauty recipes lacking, ranging from those for sunburn to prescriptions for hair which is too thick. Herbs were worn as amulets to protect the wearer not only from illness but from nightmare, lunacy and weariness when going on a journey. Betony, Vervain, Mugwort, Peony, Yarrow and Waybroad were all commonly used as amulets. Within living memory beads made of Peony roots were worn by country folk in these islands and there are still parts of the Western Highlands where Yarrow is regarded as a sacred plant.

Further, it is pleasant to realize that, in contrast to the Con-

ARTEMISIA

From "The Old English Herbals," by Eleanour Sinclair Rohde (Longmans, Green, 1922), by permission of the author.

tinental nations, our Saxon ancestors seem to have had a remarkable knowledge of plants. It has been computed that they were familiar with, and had names for, at least five hundred plants, whereas in the earliest foreign herbals not half that number are mentioned. I always regret that we have not kept more of the good old Saxon names. "Wergulu" is to me more suggestive of the fierce character of the Nettle than its present name, and I prefer "Waybroad" as a name for the flat and humble wayside "plantain."

The *Lacnunga* is another important Saxon Manuscript—a small, fat volume with no illustrations, though some of the letters are rudely ornamented. Like the *Leech Book of Bald* this MS. contains incantations suggestive of children's rhymes, notably "Nine were Nonnes sisters" a counting-out charm. There is a good deal in this MS.: about smoking not only human beings but also cattle and swine with the fumes of herbs. Many of the prescriptions end with the comforting assurance in regard to the patient: "Soon it will be well with him." Some are written in a nonsense language also suggestive of passages in children's rhymes. Thus one incantation runs, "Gonomil, orgomil, marbumil, marbsai tofeth," etc.

The oldest illustrated Saxon MS. herbal that has come down to us is the translation of the Latin *Herbarium Apuleii Platonici,* the original of which is supposed to date from the fifth century, though no copy so old as this is in existence now. This Saxon translation is ascribed to the school of Ælfric of Canterbury and was penned, therefore, some time between A.D. 1000 and 1050. Apart from the striking frontispiece of Plato, being presented with a large volume of Æsculapius and the Centaur, the book is illustrated throughout with rough pictures of both herbs and animals. There is an attractive picture of the Madonna Lily which, in spite of its botanical inaccuracy, portrays the character of the flower in a curiously fantastic manner. The stamens stand out beyond the petals and look like rays of light. To me it suggested at once Ben Jonson's matchless description of this lily—"the flower of light."

BEE-LORE

IN THE CATHOLIC LITURGY

MARTHA G. STEARNS

THE Catholic Church during her long centuries of life has perfected a ritual which is as old as she is; and in using certain material substances, and endowing them with mystical meaning, she admits Nature itself to a share in the service of the Altar, thus fulfilling the words of the Psalmist: "All ye green things upon the earth, *everything* that has life and breath, praise ye the Lord."

Light is the symbol of God Himself, and many are the prayers and psalms in which this figure is used. On Candlemas, God's Mother brings Him to His own temple, "the Light for the enlightening of the nations," and in honor of this the candles are blessed in a prayer addressed to Him "Who created all things out of nothing, and by the labor of the bees, following Thy commands, hast brought this liquor to the perfection of wax. . . ." So, because candles have a very special place in the Liturgy, the bees, living their hidden but productive community life, which is the very picture of the religious vocation, are honorable coadjutors. The candles are symbols, too, of the human body spending itself to give life to the soul, the living fire which never diminishes with all its giving of itself; and they must forever be made of the purest beeswax, that mysterious substance which the bees labor to make out of their own bodies and which has mystical light-giving properties.

Dom Theodore Bailey, O.S.B., an English Benedictine monk and student of the Liturgy, has said: "There remains but to mention the blessing of bees found in the *Rituale Romanum,* which recalls the way in which the bees may be said to assist in the celebration of the Divine Mysteries by providing wax for the candles which must burn the while. The prayer goes on to ask that 'Thy holy blessing may come down upon these bees and bee-hives; making them to increase and be fruitful, and protecting them from harm, so that their fruit may be used to Thy praise. . . .' This blessing, in the marvellously concise language of the Latin rite, sums up in a few words the place of bees in the Liturgy, as providers of the church lights, and then asks for the three things which are of most concern to the bee-keeper, namely, the increase of his colonies, the fruitfulness of their labors, and last but by no means least, protection from the many misfortunes that may befall even the most carefully tended bee-garden. Few indeed of God's creatures hold so honorable a place in the sacred Liturgy and can lay claim to so large a share in furnishing the altar for the holy Sacrifice of the Mass."

OF FRITILLARIES

ELIZABETH WADE WHITE

THERE is a sort of mystery surrounding those two ancient garden plants, the Checker Lily and the Crown Imperial, which neither old-time familiarity nor present-day neglect seem able to dispel. Why were they, considering their lack of medicinable convenience, included almost invariably in the old herbals and herb-gardens, and why are their brilliant and strange blossoms, in these days of color-seeking horticulturists, so seldom found in the springtime bed or border? The early writers are full of admiration for the tall-spiked flame-encircled *Fritillaria Corona Imperialis* and for its oddly differing little brother, *Fritillaria Meleagris*, whose delicately hung bell-like flowers seem to ring a faint and magic music which our ears, alas, are not attuned to catch but only to imagine. Parkinson[1] tells us that "The Crowne Imperiall for his stately beautifulness, deserveth the first place in this our Garden of delight"; and Gervase Markham[2] ranks it "of all flowers both Forraigne and home-bred, the delicatest and strangest." In the rich ambulatory English of the great herbals which flowered when our language was at its green summer's height, these plants are lovingly and lengthily described; and perhaps, as usual, Gerarde[3] surpasses all the others in his telling of the Crown Imperial and its indestructible "sixe drops of most

[1] *Paradisi in Sole*, 1629.
[2] *The English Husbandman*, 1613.
The Herball (Johnson's Edition), 1636.

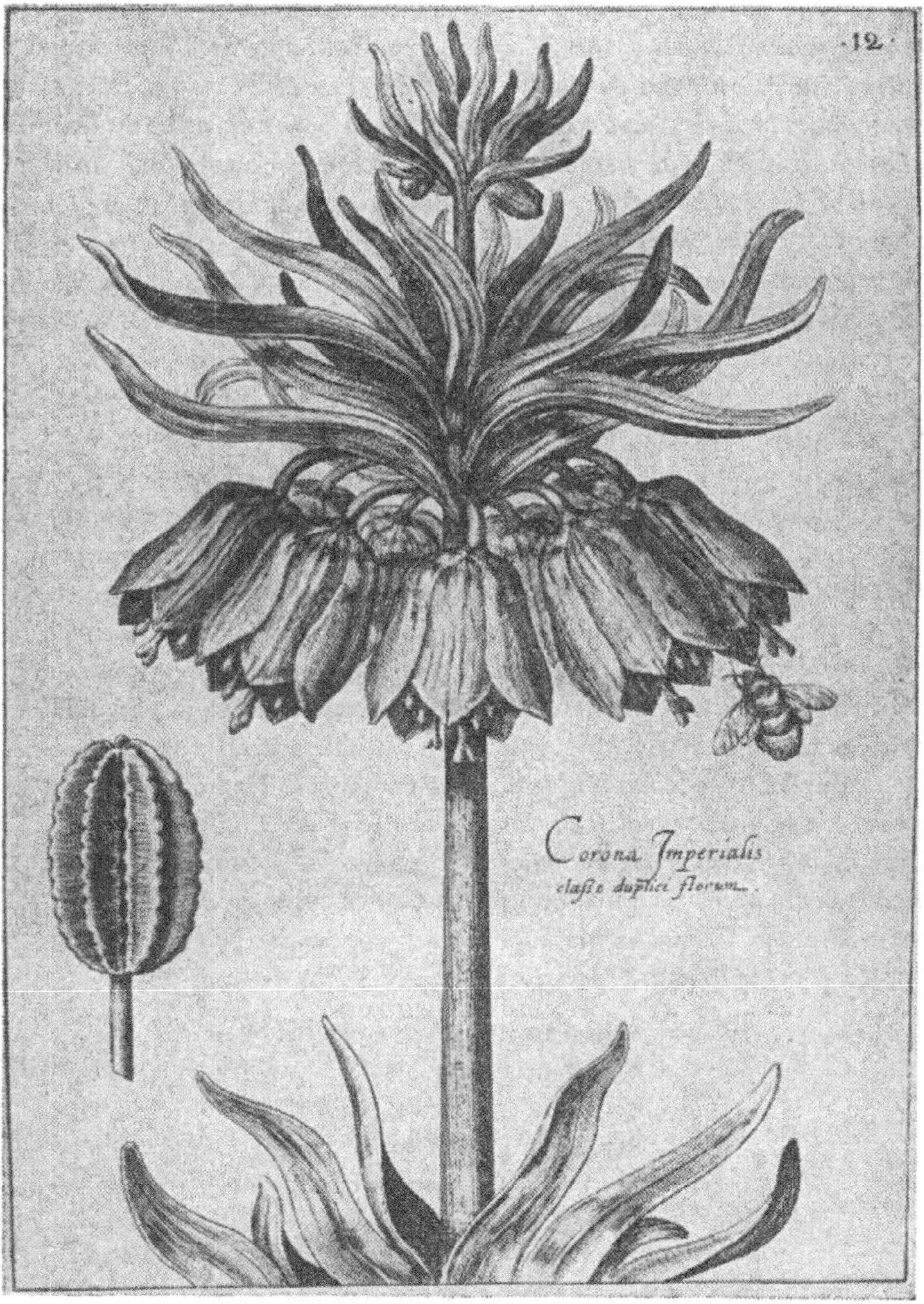

Fritillaria Corona Imperialis — The Crown Imperial
From **Crispin van de Pass — Hortus Floridus — 1614.**

clear shining sweet water, in taste like sugar, resembling in shew faire orient pearles" (which legend makes the tears of a wronged Persian queen), even as his "Checquered Daffodill, or Ginny Hen Floure" surpasses "the curiousest painting that Art can set downe; insomuch that every leaf seemeth to be the feather of a Ginny hen, whereof it tooke his name." And far from scorning this small Fritillary (which Salmon[1] called "a glorious and admirable Flower, of an extraordinary Beauty"), for its apparent lack of usefulness to the apothecary or still-room, he leaves it "greatly esteemed for the beautifying of our gardens, and the bosoms of the beautiful."

These passages speak for themselves, and for the curious flowers which with such reverence they perpetuate. And it is indeed a pleasant experience to reread them on an early Spring day, and then to go into the wind-nipped garden and find there, balancing on slender sea-green stems, the mysterious rounded downcast blooms of the checkered Fritillary, and, perhaps already pushing vigorously out of the earth, the strong green spike that promises a scarlet regal circlet, the Crown and Pearl of old cottage gardens, the Kaiser Crown of the nineteenth century, and the Crowne Imperiall of Parkinson's deathless Garden of Delight. The culture of these plants is not difficult, in fact *Fritillaria Meleagris* has naturalized itself for years in moist English meadows, and will, in tolerable soil, reappear with increase each Spring even in the most exposed New England gardens. Crown Imperial is more demanding, needing for its royal dignity a deep rich soil and a sheltered position. Once established, it will thrive, but too often it encounters such a fate as George Herbert, in his beautiful poem "Peace," describes:

> "Then went I to a garden, and did spy
> A gallant flower,
> The Crown Imperiall: Sure, said I,
> Peace at the root must dwell.
> But when I digg'd, I saw a worm devoure
> What show'd so well."

To see the bright stalk falter in its growth and become brown and

[1] *The English Herbal,* 1710.

limp, to dig it up and watch the unspeakable wire-worm at its deadly work, is a tragedy that only increases the gardener's determination to struggle on and to defeat the horticultural forces of evil at length. It is a pity that these lovely flowers are no longer enjoyed in many gardens; they have certainly, with the legends and the fame that accompany them, an historical place in the revival of herb-gardening; and for their properties, it is interesting to note (in Mrs. Grieve's *Modern Herbal*) that the bulbs of both have recently been discovered to contain poisonous qualities, and Crown Imperial (whose much-discussed foxy scent made Parkinson suspect its power) has yielded a strong toxic known as Imperialine. But the wickedness in the dark rough bulbs will, if put away safely underground, be transformed by the old earth-alchemy into a beauty of leaf and blossom which will delight and haunt the beholder, as it must have haunted Amy Lowell when she wrote, in "The Book of Hours of Sister Clotilde":

"She twined the little crocus blooms
With snowdrops and daffodils, the glooms
Of laurel leaves were interwoven
With Stars-of-Bethlehem, and cloven
Fritillaries,
Whose color varies."

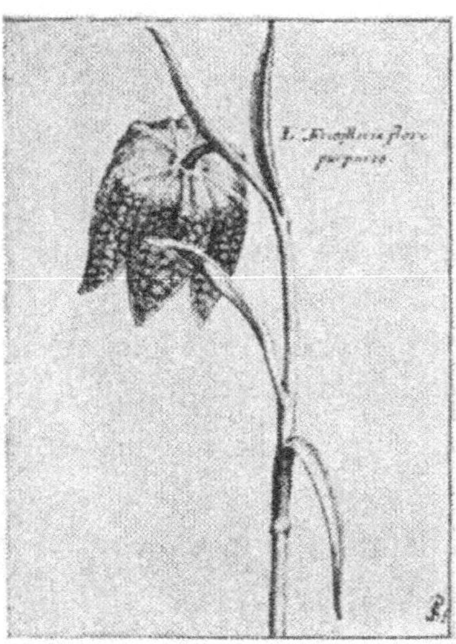

Fritillaria Meleagris — The Checker Lily
From Crispin van de Pass — Hortus Floridus — 1614.

AMERICAN DITTANY

Cunila Mariana origanoides (L.) Britton

SYNONYMA	COMMON NAMES
Satureia origanoides, Thymus	Mountain Dittany, Mint-leaves Cunila
Calaminta mariana mucronatis rigidionibus erecta Virginiana.	Wild Basil, Maryland Cunila, Stone-Mint.
Chamaedrys verna odoratissima	Sweet Horse Mint

CUNILA MARIANA is found in dry woods and rocky waste places through most of our sea-board states from New York to Georgia. It was once recommended for fever and headache and the expressed juice in milk as an application to the bites of serpents (Schoepf), but Barton writing in 1818 says, "The Dittany belongs to the class of stimulating, warm and grateful aromatics which comprises mints, monardas and pennyroyal and does not probably differ essentially in medical virtues from any of these. As a warm, grateful refreshing and slightly stimulating 'tisan it may be safely recommended."

The following excerpts from *Meehan's Monthly*, 1897-98, give an account of the plant.

American Dittany "was first made known to botanists by the

From Rafinesque's "Medical Flora"

Reverend John Bannister who in 1680 sent to the famous Ray, a catalogue of the plants of Virginia, in which he styled our plant 'Chamaedrys verna odoratissima.'

"Plukenet published numerous figures of plants in 1691 and gives a very good portrait of this as, Calaminta mariana mucronatis rigidionibus et crenatis foliis . . .

"We have Clayton sending Virginia plants to Gronovius who published in 1737, the *Flora Virginica* . . . In this it was supposed to be a member of the Thyme family and in the edition of 1762, it is removed to Satureia . . .

"Morison in his *History of Plants,* issued in 1715, describes it as 'Calaminta erecta Virginiana' and says that it is sometimes called 'Dictamnus Virginianus,' a fact already noted by Plukenet who remarks further that Dictamnus and Dittany are synonymous.

"It appears in the earlier works of Lineaus as 'Satureia ocymoides' and later as 'Melissa pulegeoides' . . .

"In Shreiber's edition the genus *Cunila* was founded for it under which it has since remained.

"It is wholly composed of American species of which there are now 17 recognized, mostly Mexican and Brazilian, our Cunila Mariana, being the most northern of all . . .

"In 1888 Mr. Pringle discovered an undescribed species of great beauty in Mexico, which Dr. Asa Gray named 'Cunila longiflora.' The name 'Cunila' is employed by Pliny and other classical writers to describe some plant of the labiate family, having spikes of flowers growing in cone-like heads; and as the bracts of the original Dittany are enlarged, enclosing the flowers, and arranged very much like a pine-cone, it is probably from this that Lineaus had the name suggested to him, seeing that the plant was already known as Dittany in our world. . . . Though not having the cone-like flowers of European Dittany, it preserved the historical connection of the popular name. . . .

"It has been stated that the volatile oil thrown off from the plant, is inflammable and that on a lighted match being applied near the stem, an explosion will follow. It is just possible that there may be some peculiarity of this kind which permits water to be drawn up through the dried stems, presenting shaving-formed ice-crystals around the stalks in the winter season and which has drawn to Dittany a large share of popular attention. A few other plants have the same power: 1. Helianthemum Canadense. 2. Pluchea bifrons. 3. Pluchea camphorata. That these ribands of ice are simply frozen matter is clear enough, and that the water is drawn up from the earth through the dead and dry stems, continually freezing on the inner edge and forcing that already frozen, toward the anterior edge of the riband, is also clear but why these few plants alone should have this hydrostatic power is the point to be solved."

References

Gron. Virg.
Schoepf. Mat. Med. Am.
Hort. Kew
Mich. Fl. Boreali Am.
Vahl enum.
Pluk. Mant.
Hist. ox.
Stokes Bot. Mat. Med.
Pursh Fl. Am. Sep.
Muhl. Cat. Pl. Am. Sep.
Elliot's Sketch
Bart. Prod. Fl. Ph.
Comp. Fl. Ph.

A CORRECTION

NOTES ON THE TRUE IDENTITY OF "OCYMUM SALINUM," THE SO-CALLED "SALINE BASIL" OF CHILE

HELEN NOYES WEBSTER

IN the article "Seven Basils," page 37 of the *Herbarist*, 1936, I quoted from Molina's *Natural History of Chile*, 1782, the paragraph describing this herb which, by the name "*Ocymum salinum*," no modern gardener, as far as I knew, had ever seen growing or dried on an herbarium sheet.

I have been puzzled and intrigued by the various references and descriptions of this South American herb which, because of its secretion of salt, seems to be of some economic importance.

After Molina's description, the Botanist Bentham, in 1832, tried to reconcile O. *salinum* with O. *minimum*.

DeCandolle, 1840, regarded "O. *salinum*" as perhaps a cultivated variety of *Ocymum basilicum* and thought "he may have seen a dried specimen in the herbarium of the London Horticultural Society and also one in cultivation."

As for me, I might still be wandering through botanical literature had not Dr. I. M. Johnston, Research Associate of The Arnold Aboretum, Harvard University, come to my rescue.

Dr. Johnston, whose special interest is the flora of temperate South

America, has called my attention to an interpretative note on the "Saline Basil" by R. A. Philippi, a German who settled in Chile about 1850 and became the greatest Chilean botanist.

I quote literally Dr. Johnston's translation of Philippi's note in Spanish, *Anal. Univ. Chile,* vol. 22, p. 703 (1863). "On the basis of Ovalle, whom he did not cite in the first edition, Molina refers to a plant which secretes salt and which he has named *Ocymum salinum.* As I have already named in these Anales (de la Universidad de Chile) this is not *Ocimum minimum,* as believed by Bentham, but rather *Frankenia Berteroana.*"

Dr. Johnston continues in his letter to me regarding the correct nomenclature of this Frankenia wrongly classed as an *Ocymum.*

"From this we can believe that Philippi had an earlier note on your name in some volume of the *Anales Univ. Chile* previous to volume 22. I have not run across it in my work, but I have not hunted for it.

"The account of the Frankenia by Alonso de Ovalle, *Histórica Relacion del Reyno de Chile,* vol. 1, chapter 13, paragraph 3 (Rome, 1646), describes a plant growing in the saline marshes at Lampa, between Santiago and Valparaiso, the region in which the Frankenia is well known.

"All Chilean botanists now agree that Ovalle's plant and *Ocymum salinum* (since Molina knew this salt plant only from the writings of Ovalle) must be the salt-secreting *Frankenia Berteroana* of central Chile.

"As I have indicated, *Contr. Gray Herb.* 70:92 (1924), Molina's specific name being older than 'Berteroana' it must be used and this Frankenia should be called *Frankenia salina* (Molina) Johnston. *Ocymum salinum* Molina is its synonym."

In conclusion the Herb Society of America is gratefully indebted to Dr. Johnston for the permission to use in its publication, the *Herbarist,* some of his information in finally establishing the identity of *Frankenia salina* and removing forever this herb from our consideration as a possible member of the genus *Ocymum.*

Sassafras Sellers

An Extract from a Letter from Mrs. John Gibbon

Asked to investigate the old Sassafras Sellers of South Street, Philadelphia, Mrs. Gibbon writes:

"Nary a trace could I find. I made myself an animated interrogation point to those who can recall hearing their parents' parents tell of the days of which Mrs. Trollope wrote.

"Those were the days when chains were stretched across the streets on Sundays—firm and stern reminders that man's horse as well as man himself should rest upon the Seventh Day. The days, too, when gentlemen took their own baskets over their arms to market, and whether to buy a fat pig or a green turtle, it was done with the solemnity and consideration due the art of the cuisine. The days when eating and drinking belonged to the fine arts—when gentlemen imported for sale fine wines and liquors 'fit for a gentleman's table.' Incidentally, Mrs. Trollope gives a charming description of sassafras. 'The sassafras is a beautiful shrub, and I cannot imagine why it has not been naturalized in England, for it has every appearance of being extremely hardy. The leaves grow in tufts, and every tuft contains leaves of five or six different forms. The fruit is singularly beautiful; it resembles in form a small acorn, and is jet black; the cup and stem looking as if it were made of red coral.'

"At my own dooryard I finally came upon a bit of information from a native son of Pennsylvania who is a 'handy man' about the place. Since I employ him, therefore let him pass for a gardener.

" 'Naw,' said he, warily shifting the forbidden quid to his off-cheek. 'I ain't never heerd tell of any sassafras sellers in Philly, but upstate where I come from there was an ole feller useter sell it. I seen him often.' Further inquiry developed that he sold thin strips tied up in small bundles for twenty cents and the residue of chips for five cents a box. 'That feller useter make $3 or $4 a day sellin' that air stuff.' With savory memories of Southern gumbo filé[1] floating through my head, I asked also ungrammatically, 'What was it used for?' 'That's what I useter wonder,' he replied. 'My mother put it in apple butter, but I ain't never heerd of usin' it for nothin' else.' A little later he appeared with a horny fistful of moist aromatic henna colored parings of sassafras! 'Where did that come from?' said I. Said he, with a grin which I felt was a shade contemptuous toward an herb grower, 'Offen that 'ar sassafras tree you got down there by the orchard fence.' He then explained that the root of the sassafras tree is sold for use. By digging down and cutting from the ends of the roots the tree is not damaged, and by making the circuit of the roots from time to time, a goodly supply can be obtained. It is the chief ingredient of root beer and sarsaparilla, is it not?

"I now flavor my apple sauce and pies with my own filé, 'eat it up and call it good.' "

[1] An article on Filé, a native American condiment, by Edgar Anderson will appear in the *Missouri Botanical Garden Bulletin*, Vol. 25, No. 3—March, 1937.

Theophrastus succeeded Aristotle in the gournment of the Schoole at Athens, about 114 Olymp. which was some 322 years before Christ. He among many other things writ a History of Plants in ten bookes, and of the causes of them, eight bookes, of the former ten there are nine come to our times reasonable perfect, but there now remain but six of the eight of the causes of Plants.

Padacius Dioſcorides Anazarbus lived in the time of Cleopatra, which was ſome few yeares before the birth of our Saviour. He writ not only of Plants: but in five bookes hath written of the neceſſarie matter of medicine.

Conventional Portraits from Old Herbals

The figures shown herewith are taken from the frontispiece of:

Botanologia

The English Herbal or History of Plants

By

Will Salmon, M.D.

Printed, M.DCC.X.

The names are reversed in the frontispiece of:

The Herball or Generall

Historie of Plantes

Gathered by John Gerarde of London

Master in

Chirvrgerie

Printed in London, Anno, 1636

Reference to "Andre Thevet's Portraits et Vies des Hommes illustre, etc., Paris, 1584," makes it clear that the portraits are rightly labelled in Salmon's Botanologia and wrongly in Gerarde's Herball. Both the above Herbals are now in the library of the "Greenman", Stonehurst, Shanty Bay, Ontario.

The first to notice this discrepancy was the "Greenman", Anno, 1909.

Old English Herbs from the Olitory at Stonehurst, Shanty Bay, Ontario, Canada.

Balm

Basil

Dill (Seeds)

Fennel

Marjoram

Mint

Nasturtium Seeds (Pickled)

Mint

Parsley

Sage

Summer Savory

Tarragon (Leaves

" (Ground)

Thyme

The Herbs furnished from the Olitory at Stonehurst, may be used dry at the table to flavour soups, sprinkle potatoes, etc., the same as you use pepper and salt.

These herbs are grown in suitable soil, in sunshine or shade as is most fitting, harvested at the right time, dried in the proper way, carefully winnowed and finely powdered. They are absolutely pure and much stronger flavoured than herbs usually obtainable.

---

1 Jar....................	.50
Box of 6 jars.............	$2.50

These herbs make an attractive and useful Christmas present. A "Herbal" of historical value and containing many receipts from the old herbals, beginning with "Dioscorides" of the first century A. D. up to the XVIII century, is sent with each box.

Toronto Office

THE GREENMAN

33 Temperance Street Elgin 2953

Keep the Memories of Yesterday
by
Planting HERBS *Today*

Our herb collection of a hundred or more varieties will meet your every requirement.

Whether they are to be used for fragrance as Thyme, Balsam or Rosemary, or for their showy flowers as Aconitum, Foxglove, Columbine, or for the sentiment they bear as Mints — Roses — Forget-me-nots or whatever is dear to your heart; — herbs for flavoring and seasoning for potpourri and perfumes; you will find nearly all of them in our booklet, "Plants for the Herb Garden."

Are you among those who have yet to see the charming effects that Germander, Winter Savory, and Hyssop provide particularly when used as edgings in trim "Knot Gardens" and perennial borders?

May we send you a copy of "Plants for the Herb Garden"?

Let us remind you, too, that we grow Evergreens, Trees, Flowering Shrubs, Vines, Roses, and Perennials in a wide variety, enabling us to provide all your garden needs. These are illustrated and described in our 96 page Handbook for 1937 — also mailed on request.

BAY STATE NURSERIES

Incorporated

NORTH ABINGTON MASS. and FRAMINGHAM MASS.

Telephone Rockland 26 — Framingham 6191

The HERBARIST

A Publication of

The Herb Society of America

No. 4

For Use and for Delight

BOSTON, MASSACHUSETTS

1938

The Herbarist for 1938
A Limited Edition
Written and Published by
The Herb Society of America
Price $1.00

Copies of the 1935, 1936 and 1937
Herbarists may be purchased
at $1.00 each.

Address
The Herb Society of America
Horticultural Hall
300 Massachusetts Avenue
Boston, Massachusetts

C O N T E N T S

*The Kings Herb-woman, & her 6. Maids, wth
Baskets of Sweet Herbs & Flowers, strewing the way.*

THE KING'S HERB-WOMAN, MARY DOWLE, AND TWO OF HER SIX MAIDS, 1685
From *Sandford's Coronation of King James II and Queen Mary*

SUGÁNDA

An Interesting Aromatic Herb

E. D. MERRILL

SUGÁNDA, more or less widely known in Spanish-American countries and in the Philippines as *orégano*, is a distinctly interesting aromatic herb perhaps deserving of more attention than it has hitherto received among those who are interested in odors and flavors in plants, and the possible use of these for culinary or other purposes. In discussing the species in 1937[1] I deliberately accepted the local Philippine name *sugánda* for the plant because the more widely used Spanish name *orégano* properly belongs with an entirely different and not closely related European herb, *Origanum vulgare*, known to us as marjoram. The latter has nothing in common with *Coleus amboinicus*, except that both are representatives of the same natural group, the mint family (*Labiatæ*), and both are markedly aromatic when crushed. It seems too highly undesirable to perpetuate the erroneous use of the name *orégano* for this particular plant.

There is, however, other than *sugánda* and the erroneous application of the term *orégano* to this species, a host of local names from which one might select, such as *pathor chur*, *pater chur*, *owa*, and *pashana bledi* in India, *bangoen-bangoen*, *daoen djinten*, *daoen koeting*, *daoen hati-hati*, *daoen kambing*, *adjeran*, *atjerang*, *madjha nereng*, *koenoe etoe*, *sukan*, and *iwak* in the Netherlands East Indies, and even *rau thom loung* and *tja to tau* in Cochinchina. From the standpoint of euphony, *sugánda* seems to be preferable to any of those listed above. Generally speaking, when one finds a considerable number of related or unrelated local names applied to a particular species in such regions as India and Malaysia, it is rather good evidence that such

[1] Merrill, E. D. Coleus amboinicus. Addisonia 20: 11-12. *pl. 646.* 1937. The illustration accompanying this paper is a black and white reproduction of this colored Addisonia plate.

plants have certain definite characters, uses, and attributes; and this is true of *Coleus amboinicus.*

From a purely botanical standpoint the plant is interesting in that it is the type of the genus *Coleus,* that is, the particular species on which the generic characters were originally based. Various forms of other species of this genus, characterized particularly by their brightly colored leaves, are commonly cultivated in homes and in gardens, but these forms with colored foliage are practically non-aromatic. The normal form of *Coleus aromaticus* has its stamens united into a sheath surrounding the style, and it was from this character that the generic name was derived; yet abnormal forms of the species occur in which the stamens are free nearly or quite to the base, therefore transcending the limits of the genus *Coleus* as normally defined, and conforming to the characters of the allied genus *Plectranthus.* The net result has been that certain botanists who happened to study the form with abnormal flowers have disposed of the species by transferring it to *Plectranthus.*

Under the binomial system this distinctly characteristic species has been blessed or cursed by a number of synonyms, the oldest name being the accepted *Coleus amboinicus* Loureiro (1790). Synonyms are *Plectranthus amboinicus* Sprengel (1825), *Plectranthus aromaticus* Roxburgh (1832), *Coleus Suganda* Blanco (1837), *Coleus aromaticus* Bentham (1831), *Coleus carnosus* Hasskarl (1842), and *Coleus suborbicularis* Zollinger and Moritzi (1854). In the pre-Linnæan literature it was described by Rheede to Drakestein under the name of *Iribeli* in 1689, and at about the same time by Rumphius (not, however, published until 1747) as *Marrubium album amboinicus;* the latter part of this name was the source of Loureiro's specific name under the binomial system, and this being the oldest binomial is the one accepted by modern botanists.

Just where this interesting plant is native cannot be stated with certainty, although some part of British India is indicated. It is an introduced plant in the Malaysian region, as it is in tropical America, for in both of the latter regions it occurs only in cultivation. It was apparently an early introduction into Malaysia by the invading colonists from India perhaps between 1500 and 2000 years ago. It is

Coleus Amboinicus
Plate 646, *Addisonia*

uncertain when it was introduced into tropical America, but simply because the Spanish name *orégano*, applied to it in the Philippines by the colonizing Spaniards, is also its name in Mexico, one surmises that the plant was one of the numerous ornamental and economic species actually introduced into Mexico through the medium of the old Acapulco-Manila galleon route. From shortly after the time Spain colonized the Philippines in 1550, this trade route was established, and persisted essentially as a government monopoly until about 1815. For this long period of time the Philippine colony was governed more or less as a dependency of New Spain, and communications between Madrid and Manila were via Vera Cruz, Mexico City and Acapulco, thence across the Pacific Ocean via Guam to Manila, there normally being one ship sent back and forth across the Pacific each year. Naturally in this long period of colonial history very numerous economic plants of American origin were introduced into the Philippines, and likewise many Old World species were introduced into tropical America from the Philippines via Acapulco.

It is claimed by all early writers that *sugánda* has certain medicinal qualities, but apparently these supposed curative properties are not to be taken too seriously. Loureiro (1790), an early botanist interested particularly in medicinal plants, states regarding its properties, "Resolvens, tonica, cephalica. Prodest in asthmate, antiqua tussi, affectionibus epilepticis et convulsivis. Odor totius plantæ vehemens, nec ingratus; sapor subardens. Usus etiam edulis." He thus notes, what Roxburgh and later authors indicated, that the plant was edible. Various opinions assembled by Sir George Watt from Indian sources indicate its use, if not its value, in the treatment of such ills as chronic dyspepsia and colic. In Malaysia it is used in the treatment of fevers and some consider it to be of value in the treatment of wounds. It is also crushed and used for giving freshly laundered clothes a pleasant odor.

Roxburgh noted that the plant was delightfully fragrant and that its leaves were frequently eaten with bread and butter, also bruised and put into country beer, cool tankards, etc., being an excellent substitute for borage; and Dalzell and Gibson stress the point that in India the plant forms an agreeable addition to cooling drinks used in the hot season. In some parts of the Malaysian region the leaves are cooked with goat flesh, whether to add a particular flavor or to disguise the not always too pleasant taste of goat meat is not stated.

Sugánda is now rather extensively used in the tropics of both hemispheres for various purposes. It commends itself to those who have proper facilities for its growth in temperate regions, for once established under proper conditions, it luxuriates with a minimum of care and attention. In 1832 Roxburgh wrote: " This most agreeable fragrant plant is common in almost every garden in India, but very rarely found in flower; this, viz. the beginning of June, is only the second time I have found it in that state in the [British East India] Company's Botanic Garden at Calcutta." Blanco made much the same observation after a very long residence in the Philippines: " Por mucho años estuve con gran deseo de ver sus flores, y nunca yo puede lograr, hasta que con el Norte de Ilocos en el Pueblo de Bintar, tuve el gusto de verlas." In nearly twenty-two years' residence in the Philippines I never saw the plant in flower. Yet in the greenhouses of the New York Botanical Garden the plant flowers freely each year.

In the tropics this species thrives in rather poor sandy soil, exposed to full sunlight. Being not only aromatic but also distinctly succulent, with thick, brittle, very fleshy leaves, it appeals not only to those who are interested in odors and flavors, but it should also appeal to those who are interested in succulent plants. It is very easily propagated by cuttings. It cannot be grown successfully in temperate regions outside of greenhouses and well-heated rooms, for the species is strictly a tropical perennial. Its requirements are, however, simple, and once proper heat and light conditions are provided, luxuriant growth results. For those who are interested in new or perhaps " exotic " flavors in food or in cooling drinks, *sugánda* is recommended as a novelty well worthy of consideration.

Courtesy of the Yenching Institute, N. Y.

EMPRESS KŌMYO (710-793) GATHERING HERBS FOR THE SICK

THE EMPRESS KŌMYO

HARRIET ADDAMS BROWN

TWELVE hundred and twenty-eight years ago, in the glorious period of "The Heavenly Peace," was born the Princess Asuka of the great Fujiwara clan. Described as a rare beauty, famous for her wit and her poetical talents, with raven tresses which reached the ground, she was chosen as the bride of the Prince Imperial, who later ascended the throne as the Emperor Shōmu. In spite of the opposition of one Court faction and in spite of precedent to the contrary, the Lady Asuka was created Empress-Consort with the title of Kōmyo, or "Refulgence." It is under this name that her memory is revered, and by it she is known to posterity and to fame.

The eighth-century culture of Japan, with its delight and proficiency in the arts, and the strong feminine and ecclesiastical dominance in politics, has been compared to the Tudor period in English history. Added to artistic receptivity, however, was a strong and united feeling of sincere religious emotion which has its counterpart in Europe in the centuries of cathedral building and in the fervor of the mediæval saints. The early Buddhism preached the virtue of humility and we find the subtle and fastidious Empress vowing, and performing her vow, to wash the feet of one thousand beggars in her devotion to the tenets of her faith.

Other ideals of early Buddhist teaching were compassion and the fellowship of all living beings: "The duty and joy of every Buddhist (that is, of every one who dwells in the all-embracing Communion of Life with conviction thereof) must be in doing every deed, speaking every word, and thinking every thought with the pious intention of dedicating his best to the profit of all."[1] These ideals were put into practical demonstration by the Empress Kōmyo, who from her own revenues erected two large buildings, one a free bath house, and one which housed a clinic where free drugs and Herbal remedies were dispensed to the sick and needy poor. Though this enlightened em-

[1] *Buddhist Art in its Relation to Buddhist Ideals*, M. Anesaki.

press died in 760 A. D., her works lived after her, and the charges of her foundations were borne by the government for many years. In the year 825, the support was revived by imperial order, and the subsidy was continued for almost eight hundred years.

On July 22, 726, Kōmyo presented twenty-one chests of medicines to the Temple, accompanied by a meticulous list of the names and quantity of each drug. These names are inscribed in the beautiful calligraphy of the times on a scroll of white hemp paper still wrapped on its scented core of sandalwood. Her prayer for their successful use follows in the Deed of Gift.

> "We humbly beseech that all who partake of these medicines be freed from all sickness and saved from all ailments. May every thing good for them be accomplished and the evil checked and driven away, and their course of long life be not prematurely interrupted through the inevitable effect of cause beyond their control, and may they, after a natural death, be born in the eternally flowering Paradise, face to face with Vairocana, and themselves attain Buddhahood."

The inviolate archives of the Shōso-in, the Imperial Repository at Nara, preserve to this day twenty-six samples of these actual medicines which have been identified according to European terminology by modern chemical and botanical research. One enthusiastic commentator claims that "these ancient herbs have lost little of their efficacy." To this, no modern herb-user, firm in the conviction that freshness is all, could agree; still there is a fascination in knowing that these roots and barks and teas were a part of the generous legacy of that kindly and noble lady who was mindful of the least of these.

For a clue to other remedies in use at this time, we may turn to the records of contemporary China with a certain degree of security, as the close connection of the Island Empire with the Tang culture in the Arts and in Religion, presupposes an equal familiarity in the field of Medicine.

It is not surprising to read of Chaulmoogra, Ginger, Castor-Oil, Tribulis, and Ephedrine, but to come upon our familiar treasures of Rosemary, Fennel, Coriander, Carthamis, Dandelion, Angelica, Aconite, Artemisia, Verbena, Plantain, and Marigold, gives us a sudden throb of kinship with that distant land and for the gracious Empress who aided her people with the healing herbs.

JUICHIMEN KWANNON, DEITY OF MERCY

Said to have been modeled by the Abbot Rōben from the figure of the Empress Kōmyo as seen reflected in the lotus pool

Courtesy of the Museum of Fine Arts, Boston

Omitting from the list the minerals, salts and strange fossil bones of "Dragons," we meet many familiar plants in this old catalogue.

KANZO (*Glycyrrhiza glabra* L.)—Licorice, "Honey Herb," "Honey Sweet." "Old Man of the Empire."

NIN-JIN (*Panax ginseng*)—Ginseng. The indispensable restorer.

CHIKUSETSU NIN-JIN (*Panax repens* Max.).

DAIWO (*Rheum* off. Bail.)—Rhubarb, "whose root has greater strength when it is old."

KEISHIN (*Cinnomomum cassia*)—Cinnamon, for which The Charitable Medicinal Institution "petitioned 100 kin" in 759 A. D.

HIHATSU (*Piper Longum*)

SHOZO (*Polygala Tenuifolia* Willd.)
ENJI (*Polygala Tenuifolia* Willd.) —used for plasters.

GENKWA, flowers of (*Daphne Genkwa*)—the headache flower.

HAZU, seeds of (*Croton Tiglium*)—the physic nut.

KOTORITSU, sap of (*Populus Euphratica* L.) or (*Liquidamber Formosana*)

NYAKU, root of (*Lindera strychnifolia* Vill.)—for fevers.

SUHO (*Cæsalpinia sappan* L.)—Sappan wood, an astringent.

SHOMOKKO (*Aristolochia Kempferi* Willd.)—the Japanese birthwort.

CHOKO (*Eugenia aromatica* Bail.)—Clove.

BIBLIOGRAPHY

Botanicum Chinensis. Vol. II, III. 1895. E. Bretschneider.

History of European Botanical Discoveries in China. 1898. E. Bretschneider.

Science Papers. Notes on Chinese Materia Medica. Daniel Hanbury, F.R.S.

Middle Kingdom. Vol. I. S. Wells Williams.

General History of China. P. Du Halde.

Journal Asiatique. Oct.-Nov. 1852. Index of Plants of Japan and China. Hoffmann and Schultes.

The Shoso-in. 1930. Sir Percival David, Bart.

Shoso-in. A. Matsuoka. Japan Society of New York.

Catalogue of the Imperial Treasures in the Shōso-in. Published by The Imperial Household Museum, Tokyo. Text by Prof. Y. Oshima. Translator, J. Harada.

Shokubutsu Jukan. T. Makino.

Contribution Toward the Materia Medica of China. 1871. F. Porter Smith.

Catalogue of the Plants in the Tokyo Museum. Imperial University Botanic Garden.

MINTS AND MICROSCOPES

A foreword to the article by Mabel Ruttle

EDGAR ANDERSON

IT IS difficult to grow mints without losing faith in botanists. No amateur expects to be able to name all the plants of his garden, accurately and well, but he does expect that a professional botanist should be able to do so. Yet if he submits his mint collection for determination the results are almost inevitably disappointing. If the work is done honestly some of the mints will be left quite unlabelled and other names will be followed by a questionmark. Our amateur will be even more disturbed if he consults a second botanist and then finds, as he certainly will, that the two sets of names do not agree.

Yet there is here no real need for losing faith in botanists or in botany. Rather there is needed a better understanding about the whole business of plant naming. Strange as it may seem, the naming of plants is not a finished process, done once and for all time. It is rather a series of approximations which come closer and closer to the truth as botany progresses. Some groups of plants are relatively simple to provide with names, as for instance, the maples of New England. Sugar Maple, Red Maple, Mountain Maple, and Striped Maple; each one is distinct in leaf and twig and bark. But there are groups of plants which are far from this simplicity and the mints of Central Europe are such a group.

Had man taken no interest in mints they would have been somewhat difficult to classify; his love for them has tangled the original snarl. For centuries he has been digging them up and carrying them from place to place, growing them in his gardens and letting them run wild again. Strange, garden-bred hybrids which would never have occurred in nature have been spread wide by his efforts.

But man always learns at last to deal with the complexities he creates and the mints are no exception. Within the last two decades botanists have hit upon a strange new way of untangling the relationships of certain groups of plants. Root tips are imbedded in tiny chunks of paraffin and sliced thinner than the thinnest tissue paper, for examination under the microscope. From such material we learn the number of hereditary units or chromosomes which characterize each variety and form. The information as to chromosome number may be very helpful in understanding the relations between the various species of a genus. Some are simple and have only two sets of chromosomes. Such species in the genus Mentha are *M. rotundifolia* and *M. longifolia,* each of which has two sets of 12 chromosomes. Other species and varieties have four sets, as for instance, *M. spicata.* There is good reason to believe that the species and varieties with the higher numbers have come from the fusion of two or more simple ones. This new technique is still more or less in the experimental stage, but it has already advanced to the point where it may be very helpful in interpreting a difficult genus.

Dr. Mabel Ruttle (Mrs. Nebel) has been one of the pioneers in this technique and fortunately for us she turned her attention to the wild and cultivated mints. She collected and studied them as they grew wild in Europe and America. She cultivated them in her experimental garden, and many of them she submitted to detailed microscopic examination. Her work is so new that as yet even few professional botanists are aware of it. The Herb Society of America is fortunate to be able to present a discussion of our garden mints which is of practical value to the herb gardener and which at the same time is a report from the advancing front of science.

SOME COMMON MINTS AND THEIR HYBRIDS

MABEL L. RUTTLE

A GENUS such as Mentha may be regarded systematically as consisting of a very large number of small species (Trautmann, 1925, in Javorka, Magyar Flora lists 210 small species) or as consisting of a small number of inclusive species, each consisting of several subspecies and numerous varieties (Briquet, 1891, 1896; Topitz, 1913; Schinz and Keller, 1914; Hegi, 1914; Fraser, 1926, and others). If, as in this genus, the species and varieties hybridize freely in nature, a wild population is obtained with a large number of intergrading forms.

In 1928 I attempted to check with the aid of cytology, embryology, and genetics the validity of these systems of classification in Mentha. For this purpose I collected mints from different botanical gardens in Germany, Switzerland and England and in the field in the neighborhood of Berlin and Freiburg, Germany, and along the coast near Dartmouth, England. Part of this collection was brought to Geneva, N. Y., in 1929. Here the collection was further augmented by the addition of mints collected around Geneva and received from various sources. Self- and open-pollinated progenies of the species and progenies of the fertile crosses between the species have been grown. The chromosome number has been counted in root tips of the principal species and in some hybrids; the shape and size of the embryo sacs have been measured and compared.

The following plants have been studied in some detail: 1. Corsican Mint (*Mentha Requienii* Benth.), a good species, very uniform progeny; 2. Pennyroyal (*M. Pulegium* L.), a good species with diploid and tetraploid forms and hybrids between the two; 3. Woolly Mint (*M. rotundifolia* (L.) Huds.); 4. European Horse Mint (*M. longifolia* (L.) Huds.); 5. Water Mint (*M. aquatica* L.); 6. Field Mint (*M. arvensis* L.). Numbers 3, 4, 5, and 6 are all good

FIG. 1. *M. Pulegium* L.

FIG. 2. *M. longifolia* (L.) Huds.

Figs. 1-6 from *Illustrations of the British Flora*, L. Reeve & Co., pp. 195-197

species but they all hybridize readily; 7. Spearmint (*M. spicata*), probably of hybrid origin, offspring very variable. It hybridizes readily with numbers 3, 4, 5, and 6; 8. Egyptian Mint (*M. Niliaca* Jacq. em. Briq.), a hybrid between *M. rotundifolia* and *M. longifolia* and between *M. rotundifolia* and *M. spicata;* 9. Peppermint (*M. piperita* (L.) Huds.), a hybrid between *M. aquatica* and *M. spicata* or more probably between *M. aquatica* and the hybrid between *M. spicata* and *M. rotundifolia;* 10. American Wild Mint (*M. arvensis* var. *canadensis* (L.) Briq.), probably a good species, often considered a variety of Field Mint; and 11. Japanese Peppermint (*M. arvensis* var. *piperascens* Briq.), usually considered a variety of Field Mint but more probably a distinct species.

Mentha Requienii Benth. (Corsican Mint) is a small creeping mint native to Corsica and Sardinia. The refreshing minty odor of the leaves is attractive. It forms a soft green fragrant mat in the rock garden near a tiny stream where it is covered with soft purplish flowers. It is rather delicate and is mostly winter-killed at Geneva. Its seeds are fertile and the progeny very uniform. The chromosome number in the body cells is 18 (fig. 8), which sets it apart from the other species of the group.

Mentha Pulegium L. (Pennyroyal) is also an odd member of the genus *Mentha* and quite unrelated to the other species (fig. 1). I found it growing abundantly all along the coast at Dartmouth apparently escaped from cultivation. The material I used for study was growing in the Botanical Garden, Berlin-Dahlem. There were two forms growing together and not separable by appearance only; one had 20 ± chromosomes, one 40 ± chromosomes. The two forms hybridized freely, producing a population with 30 chromosomes. Seedlings of these varied greatly in size, shape and vigor. Some dwarf forms of these seedlings were more suitable for the rock garden than is typical *M. Pulegium*. This species like *M. Requienii* does not hybridize with other mints.

The large interrelated group of mints found throughout Europe, Asia Minor and North Africa has been divided by Briquet (1891, 1895), Hegi (1914) and others into five main species (figs. 2, 3, 4, 5, and 6): *Mentha rotundifolia* (L.) Huds. (Woolly Mint); *M. longi-*

FIG. 3. *M. rotundifolia* (L.) Huds.

FIG. 4. *M. spicata* (L.) Huds.

folia (L.) Huds. (European Horse Mint); *M. spicata* (L.) Huds. (Spearmint); *M. aquatica* L. (Water Mint) and *M. arvensis* L. (Field Mint). *M. rotundifolia, M. longifolia* and *M. spicata* bear their flowers in spikes and have leaves with no or very short petioles. In *M. aquatica* the flowers form a terminal head with one or two axillary clusters just below the head and the leaves have distinct petioles. In *M. arvensis* the flowers are borne in axillary clusters, the leaves are petiolate.

Each of these five species consists of several subspecies and numerous varieties. Each of the five hybridizes freely in nature with the other four. These hybrids occur naturally, some frequently and some quite rarely, and have been given specific names (fig. 7). The large number of varieties of the parent species makes for an even larger number of varieties of each hybrid. Most of the hybrids are sterile, some are partially fertile and set few seed and others are completely fertile and set abundant seed. The partially fertile and completely fertile hybrids when self-pollinated produce plants with a random segregation of the parent characters. These may back-cross to the parents and cross with each other and with other species producing still other types, so that it soon becomes impossible to tell the exact parentage of many of the plants found in nature.

Certain clones of these species and hybrids have proved of value for their essential oils. These have been cultivated in the old civilizations, certainly in ancient Egypt, and have followed migrations. Thus, in addition to natural varieties there are horticultural varieties, the production of single clones. If these were apples they would be designated by such names as McIntosh or Cortland, but being mints, they are designated by specific and varietal names which cover not only the special clones of commercial value but a large number of other clones.

Many of these species, varieties, hybrids and special clones have been brought to America. They have run wild both by seed and by clone until every little brook and glen has its mint variety, often varying just a little from that in the next glen. It also seems probable that natural crosses have occurred between these European mints and the native American mint.

FIG. 5. *M. aquatica* L.

FIG. 6. *M. arvensis* L.

M. rotundifolia (fig. 2), *M. longifolia* (fig. 3) and their hybrid *M. Niliaca* Jacq. em. Briq. (Egyptian Mint) proved easy to collect in Black Forest rambles. The branches of *M. rotundifolia* tend to stand at rather large angles to the main stem giving the plant a somewhat rigid appearance. The hairs on the backs of the deeply-veined, rounded leaves are long and branched, giving the characteristic woolly covering. In *M. longifolia* the angle of branching is more acute, the leaves longer, the hairs thick but unbranched.

The chromosome number in body cells of both *M. rotundifolia* and *M. longifolia* is 24 (figs. 10 and 9). Progenies from self-pollinated seed of both species are fairly true to type. The two species hybridize freely and produce the fertile hybrid *M. Niliaca*, also with 24 chromosomes (fig. 13). Segregation occurs in hybrid seedling populations according to Mendelian principles, crosses occur within the segregants and back-crosses to the parents with the result that *niliaca* types may be collected, in the wild, varying all the way from plants closely resembling *M. rotundifolia* to plants closely resembling *M. longifolia*. Now when one considers that both *M. rotundifolia* and *M. longifolia* are composed of numerous varieties which cross freely between themselves and between varieties of the opposing species, one can begin to form an idea of the complicated picture presented by these two species alone.

The species most closely related to these two is undoubtedly *M. spicata* (fig. 4) with a chromosome number of 48 ± (figs. 11 and 12). Certain characters of *M. spicata* resemble *M. rotundifolia*, certain others *M. longifolia*, while others are different from either. Hybrids between *M. spicata* and *M. rotundifolia* occur and are not always distinguishable from those between *M. longifolia* and *M. rotundifolia*. The chromosome number in different *spicata* seedlings varied somewhat from the expected, plants with more or less than 48 chromosomes occurring quite frequently. Sterile seedlings, very hairy seedlings, typical fertile *spicata* seedlings occurred in progenies of selfed *spicata* plants. The odor of different plants is very different, that of some being pleasant, that of others quite obnoxious. All this indicates that the species group *M. spicata* possibly represents a segregant of a hybrid between *M. rotundifolia* and *M. longifolia* in which the chromosome number has doubled but is for some reason not quite balanced.

Spearmint of commerce, commonly called "*M. spicata*," has 36 chromosomes. The leaf venation resembles *M. rotundifolia*, the clone is quite sterile and rarely or never sets seeds. It possibly represents a hybrid between a particular *M. spicata* and *M. rotundifolia* or *M. Niliaca*.

M. aquatica (Water Mint) is a large vigorous mint found in low places throughout Germany (fig. 5). I collected it also along the coast near Dartmouth, England. The flowers are borne in heads, with perhaps one or two axillary clusters, the leaves have long petioles, the odor is indifferent. The chromosome number is 96 (fig. 14).

In spite of the great difference in chromosome number I was able to hybridize *M. aquatica* and the other mint species in the 12-ploid series. These hybrids were in all cases sterile. Hybrids of *M. spicata* x *M. aquatica* are regarded as *M. piperita* (L.) Huds. This name has been applied in particular to the commercially grown "Mitcham peppermint,"[1] but it is also applied to the large group of other hybrids of similar parentage just as the name "*M. spicata*" has been applied both to commercial spearmint and to the large and variable *spicata* group.

The hybrids between *M. spicata* and *M. aquatica* which I have made and grown are distinctly poor in both quantity and quality of oil. Apparently the parents must be more carefully selected in order to produce typical peppermint or it is possible that peppermint is the product of a cross between one of the numerous varieties of *M. Niliaca* (*M. spicata* x *M. rotundifolia*) and *M. aquatica*. This possibility is also indicated by the chromosome number (fig. 16).

Besides Mitcham peppermint there are several other varieties grown which yield a good quality of peppermint oil. It is possible that the early colonists brought some of these to America as well as Mitcham peppermint. They have apparently escaped from cultivation, occasionally have set seed and so account for the various peppermints found growing wild in the various glens. Although Mitcham pep-

[1] For commercial spearmint and peppermint see Fig. 3 and Fig. 2. U. S. Dept. of Agric. *Farmers' Bulletin* No. 1555.

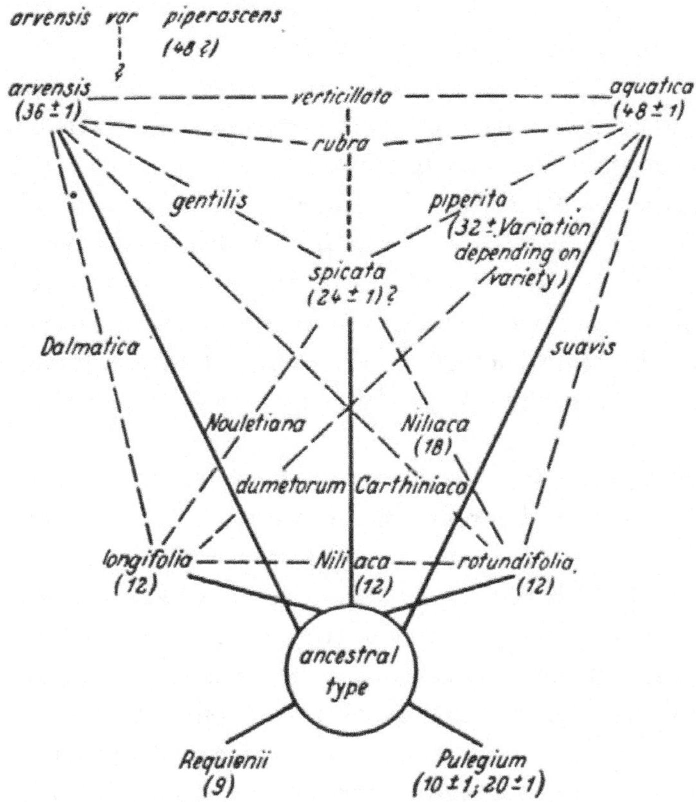

FIG. 7

[*From Ruttle, Gartenbauwissenschaft* 4, *Diag.* 1. *p.* 463]

This figure shows the five interrelated species and their hybrids indicated by the broken lines. Heavy black lines indicate the "supposed origin" of the species. As indicated M. Requienii and M. Pulegium show different lines of origin. The position of the species on the chart is arbitrary except as influenced by external morphology and chromosome number. The chromosome numbers given are half the number in the somatic cells and are consequently half the number given for the species in the text and in figs. 8-16.

permint is usually completely sterile it may on rare occasions produce anthers with apparently viable pollen and may through very occasional seeds in the course of years have yielded occasional progeny. M. *piperita* var. *citrata* Briq. (Bergamot Mint) represents a possible segregant more like the *aquatica* parent, or possibly even a mutant of M. *aquatica* itself.

The hybrid between M. *aquatica* and M. *rotundifolia* produced interesting offspring, the odor of which is distinctly pleasing. Some readers may care to grow this in their gardens.

M. arvensis (Field Mint) is commonly found along every little brook in south Germany. It is a small low-growing creeping mint with considerable variation in plant type and a most vile odor. The chromosome number is 72 (fig. 15).

The American Wild Mint, *M. arvensis* var. *canadensis* (L.) Briq. has the axillary flowers of *M. arvensis*. The habit of growth is quite different, however, and the odor is much more pleasant. This plant is often listed as a valid species under the name *M. canadensis* L. which may be more correct. The number of its chromosomes and its crossing relationships with the other species should be first determined.

Japanese Peppermint, *M. arvensis* var. *piperascens* Briq., is a tall vigorous plant very different from the small creeping field mint of the German brooks. Its self-pollinated progeny is variable as regards odor, vigor, leaf shape and resistance to mildew. Its chromosome number approaches 96. This evidently should be regarded as a distinct species. This mint is valuable commercially because of its high oil yield and the high menthol content of its oil. The roots are, however, winter-killed in New York and Michigan. To produce a plant with a good oil yield and high menthol content but with more vigorous roots this plant was crossed with *M. aquatica, M. arvensis* (European), *M. rotundifolia, M. longifolia* and *M. spicata.*

The progeny of the cross with *M. arvensis* (Field Mint) were distinctly bad in odor, the cross with *M. longifolia* was not particularly promising. The cross *M. spicata* x *M. arvensis* var. *piperascens* yielded several vigorous progeny. One plant in particular stands tall and straight, has dark red stems with long purplish spikes of flowers, large for a mint. The odor is also quite pleasant and rich in menthol. It blooms late and should, I think, be a pleasant addition to an herb garden. Another plant from a similar cross but using a different *spicata* parent has white stems and less beautiful flower spikes but a somewhat more pleasant odor. Both of these plants have distinctly vigorous rootstocks and can be grown on high land.

Among the plants of the cross *M. aquatica* x *M. arvensis* var. *piperascens* are plants with less vigorous roots than *M. aquatica* but more vigorous than *M. arvensis* var. *piperascens.* The menthol content is

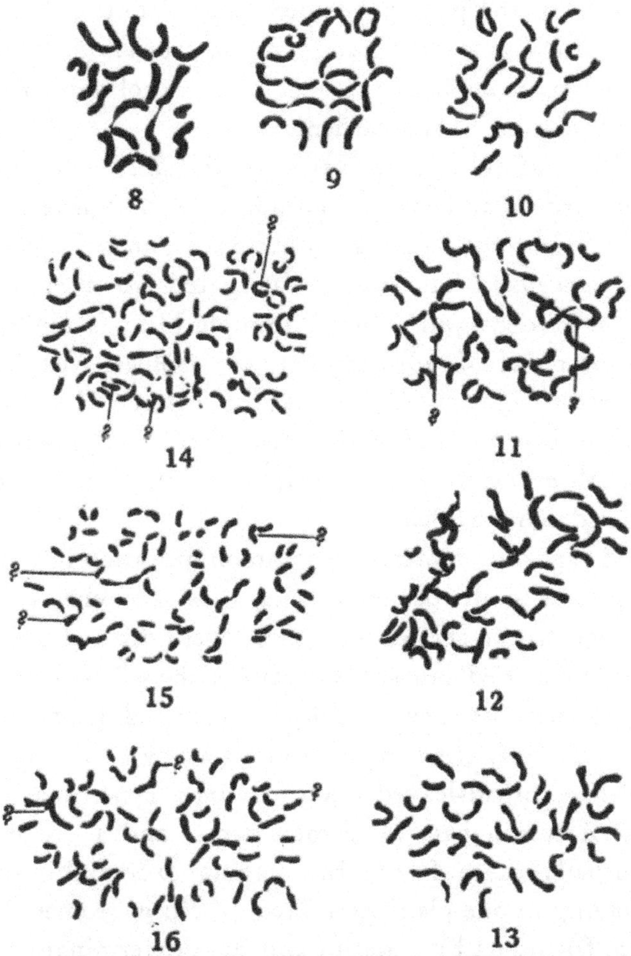

From Ruttle. Gartenbauwissenschaft 4, figs. 54-62

The number of chromosomes* in body cells of the different Mentha species. Interrogation marks indicate points at which the interpretation is somewhat doubtful.

Fig. 8. *M. Requienii.* Metaphase† showing 18 chromosomes.

Fig. 9. *M. longifolia.* Metaphase showing 24 chromosomes.

Fig. 10. *M. rotundifolia.* Metaphase showing 24 chromosomes.

Fig. 11. *M. spicata,* seedling. Metaphase showing 48 $\pm$ 2 chromosomes.

Fig. 12. *M. spicata?* Hairy type growing in *M. spicata* plot, Berlin-Dahlem. Metaphase showing 48 chromosomes.

Fig. 13. *M. Niliaca* var. *Lamarckii.* Metaphase showing 36 chromosomes.

Fig. 14. *M. aquatica.* Metaphase showing 96 chromosomes, 91 from one section and 5 from the adjoining.

Fig. 15. *M. arvensis.* Seedling. Seed obtained from Czechoslovakia. Metaphase showing 72 chromosomes.

Fig. 16. *M. piperita* "Mitcham," from the Karl Pabst Seed Co., Erfurt. Metaphase showing 65-69 chromosomes depending on whether certain chromosomes are one long or two shorter chromosomes.

* Chromosome. Formed bodies in the cell nucleus, made of protoplasm, bearers of hereditary units, constant in number and shape for each species, visible during nuclear division and easy to count at the stage of division called metaphase.

† Metaphase. The stage of nuclear division during which the chromosomes are arranged on a central cell plate and during which the chromosomes divide and the halves begin to migrate to opposite ends of the cell.

at least as high as in the parent Japanese peppermint. These plants are not so ornamental as those of the cross *M. spicata* x *M. arvensis* var. *piperascens* but if one likes the odor of menthol one will like to have a little of this mint in the garden.

Certain varieties of mint in addition to the ones described have been grown in gardens and given common names. Variegation is a common mutation in mint seedlings. I obtained numerous variegated seedlings of *M. arvensis*. The pretty variegated woolly mint commonly grown in gardens and called Pineapple Mint is undoubtedly a variegated mutant of *M. rotundifolia;* the Golden Apple Mint a variegated mutant of *M. gentilis*, the hybrid between *M. spicata* and *M. arvensis*. One suspects that in this case the *M. arvensis* may possibly have been *M. arvensis* var. *canadensis*. A count of the chromosomes might yield some evidence.

Crispa-leaved types also represent a common mutation in the mints, a mutation common to all species and many of the hybrids. Of the 2,000 seedlings grown at the Station many were crispa-leaved. The degree of crinkling varied greatly, sometimes the edges were simply deeply notched, sometimes they were deeply lacerated. Leaf size, plant type, hairiness and odor were very variable. Two types are commonly grown in gardens, a broad-leaved smooth *Niliaca* type usually sterile and one plant of which had 36 chromosomes, and a crispa-leaved *spicata* type usually quite fertile but having exceedingly variable progeny and having in one plant examined 48 chromosomes.

The evidence furnished by genetics and by a determination of the chromosome number has accordingly established the mint species whose chromosome numbers are some multiple of 12 as being a closely related group. At the same time Japanese peppermint (*M. arvensis* var. *piperascens*) should possibly be regarded as a distinct species. *M. spicata* represents a hybrid in which the chromosome number has doubled but in which certain chromosome groups are unbalanced. Spearmint of commerce is possibly a hybrid between particular *spicata* and *rotundifolia* or *niliaca* parents.

The amount of work so far done on this genus has shown mere indications of what may be done towards a further systematic understanding and classification and also towards a selection of plants for breeding for the production of plants with a higher yield of oil, with better quality of oil, with higher resistance to certain diseases, and with more vigorous rootstocks adapted to different soils. Moreover, what has been done in this genus has not yet been begun in America in the majority of the herb plants, the perfume producing plants and the drug plants.

Bibliography

Briquet, John.

1891. *Les Labiées des Alpes maritimes.* Mat. p. servir a l'hist. d. l. flore des Alp. mar. (Burnat).

1895. *Labiatæ.* In Engler u. Prantl, Natürl. Pflanzenfam. 4, 181-384.

Fraser, John. 1926. *Menthæ Britannicæ.* Report of the Bot. Exchange Club for 1926, 213-247.

Hegi, S. 1914. *Illustrierte Flora von Mitteleuropa.* Pub. by J. F. Lehmans, München, Mentha 5 (4) 2335-2357.

Ruttle, Mabel L.

1931. *Cytological and embryological studies on the genus Mentha.* Gartenbauwissenschaft 4: 428-469.

1932. *Chromosome number, embryology and inheritance in the genus Lycopus.* Gartenbauwissenschaft 7: 154-178.

1930. *The bearing of cytological studies on breeding possibilities in the genus Mentha.* Amer. Ass. Hort. Soc. 1931: 335-336.

1932. *Interspecific hybridization in the genus Mentha.* Proc. 6th Int. Cong. Genetics, vol. 2: 311-312.

Schinz, H., u. R. Keller. 1914. *Flora der Schweiz.* Publ. by Rausteen, Zurich: Mentha, part 2: 241-250.

Schnarf, K. 1917. *Beiträge zur Kenntnis der Samenentwicklung der Labiaten.* Denkschr. Akad. Wiss. Wien. 94, 211-275.

Souèges, R. *Recherches sur l'embryogénie des Labiées.* Bull. Soc. bot. France. 68: 441-464.

Topitz, Anton. 1913. *Beiträge zur Kenntnis der Menthenflora von Mitteleuropa.* Beihefte Bot. Zbl. 30: 138-264.

Trautmann, R. 1925. *In Javorka:* Magyar Flora III, Mentha: 905-970.

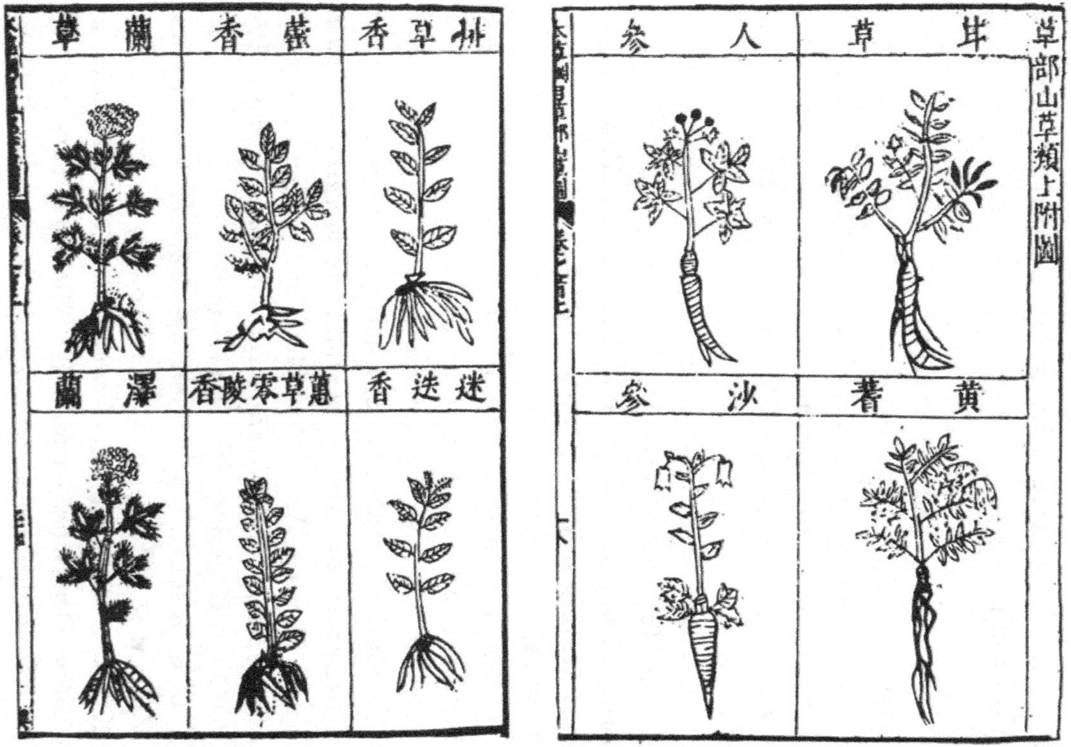

1 2 3 7 8

4 5 6 9 10

KEY TO WOODCUTS OF PLANTS FROM THE PEN TS'AO OF LI CHIH CHIEN. 1596
Courtesy of the Boston Medical Library

1. *Eupatorium chinensis* 2. *Sophanthus rugosus* 3. *Unidentified* 4. *Eupatorium Japonicum* 5. *Ocimum Basilicum* 6. *Rosmarinus officinalis* 7. *Panax Ginseng* Meyer 8. *Glycyrrhiza glabra* L. 9. *Adenophora polymorpha* Ledib. 10. *Astragalus Henryi*

Translations of the Chinese character and the botanical identification through the kindness of Miss Hirano and Mr. Robert T. Paine. Boston Museum of Fine Arts.

THE HERBAL OF LI CHIH CHIEN

An Extract of the *Pen ts'ao cang mou,* that is, *The Chinese Herbal:* or *The Natural History of China for the use of Medicine*

> This work was undertaken and composed by a Doctor of the Family or Dynasty of Ming, called Li che tchin: But Death preventing this Author from putting the last Hand to it, his Son, after having revised and augmented it, presented a petition to the Emperor Van lie the 24th year of his Reign, in consequence of which the Emperor gave Orders to the Tribunal of Rites to publish the Work, which has been reprinted the 22nd year of the Reign of the late Emperor Cang hi.
>
> R. BROOKES MDCCLI
> From the French of P. Du Halde

The Text

There are seasons proper to gather and prepare Medicinal things, some of which are to be dried in the Sun, and others in the Shade.

Commentary

Hong King says, that the time of gathering things that enter into the Composition of Remedies is the beginning of the year, and that this Custom has been established ever since the Reign of Han: The reason why the greatest part of the Medicinal Roots are gathered in the second and eighth Moon is, that in the beginning of the Spring, the Sap, rising in great abundance is in its full strength, and then only beginning to cause the Plants to bud, it is not yet dispersed nor consumed as it is afterward in the Production of the Branches and Leaves: As for the Autumnal Season the Leaves and Branches being withered, then the Juice or Sap falling downwards returns to its origin.

In short if Roots are gathered in the Spring it should generally be done in the Morning, and if in the Autumn it should be in the Evening for the same reason.

As for the time proper for the gathering Flowers, Fruits, Leaves, and the Stalks or Trunks of Plants, this general Rule is to be observed, which is, that they should be cropt in their perfect Maturity.

The King's Herb-Woman, Miss Fellowes, and Her Maids, 1821
From Nayler's Coronation of George IV

THE KING'S HERB-WOMAN

ELIZABETH WADE WHITE

IN a neglected lumber-room in Limbo there must surely repose — among tarnished pieces of armour, dusty ostrich plumes, falcon-hoods whose proud wearers long ago found lasting darkness, and other oddly-assorted, once splendid, now pathetic, trappings of ceremonial — a small gilded basket to whose faded satin lining cling a few leaves and petals, dry as dust now, but still remembering with faint fragrance the English gardens from which they came. This was the symbol of office of the King's Herb-woman, that proud figure who trod, strewing health and sweetness underfoot, at the head of how many a Coronation procession through England's centuries of royal pageantry. She is gone now from the ceremony of Coronation, along with the King's Banquet in Westminster Hall, with the mounted and armoured King's Champion who appeared at the Banquet, and with the famous foot procession between Westminster Hall and the Abbey. For changing times bring change to the oldest of customs, and not since the last and most lavish of the " Walking Coronations," George IV's, have these ancient scenes been included in the drama of the King's crowning.

The King and Queen of England who were crowned on May 12,

1937, came to the Abbey in the great golden State Coach, so there was no need for the casting of herbs and flowers along their path, and consequently no Herb-woman appeared in their procession. But the greenness of her memory, as a favorite actor in the old pageants, is shown by the mention given her in last year's crop of Coronation literature, and by the inclusion of her charming picture in a series of Coronation cigarette-cards issued by the enterprising tobacconists, Messrs. Churchman of Ipswich. So it is fitting that, before the Coronation year closes, these pages should pay some tribute to so ancient and picturesque a symbol of the craft of the herbalist.

Mystery surrounds the first appearance of the King's Herb-woman in the Coronation procession, for accounts of the crownings of the early monarchs of Britain are fragmentary in the extreme. But the reasons for her inclusion are probably twofold: an antique symbolism melting gradually into a decorative practicality. Is it not permissible to conjecture that, in the time of the Saxon kings, the old Druidical and Roman rites of earth-worship had carried over into the early Christian ceremony in the shape of a stately white-robed female figure, bearing, like Ceres or Pomona, the harvest-horn of plenty, and strewing its riches before the warrior-chieftain who had subdued his foes and won a crown? However, there appears to be no definite mention of the King's Herb-woman as an individual until as late as the middle of the seventeenth century, though then she is taken for granted as an anciently established member of the royal retinue. So we must accept the fact that whatever unearthly dignity may have ennobled the earliest Herb-women was soon replaced by the more homely and material distinction belonging to those whose duty was the purefying of the air that surrounded the King's person. From the time of King Stephen even into the nineteenth century the strewing of sweet herbs and clean straw and rushes over the floors of houses and public buildings was considered a necessity for the preserving of health, and bunches of contagion-preventing herbs were carried by physicians, supplied to judges at criminal trials, and sold by thousands during times of plague. It is not surprising, therefore, that the King's foot procession to the Abbey, whether it passed along the street itself or on a raised platform, was led by a group of women who strewed aromatic herbs to counteract germs and odours arising from the habitually insanitary public ways.

Elizabeth Burgess, Provider of Flowers and Sweet Herbs for the court of King Charles I, is the first Herb-woman whose name I have been able to discover. Though she is not mentioned as having taken part in the Coronation procession, it is probable that she holds a greater, even an unique, distinction among Herb-women: that of having given her life for her King. Her story is told with tantalizing brevity in the Calendar of State Papers, Domestic, 1660-1, where her daughter, Bridget Rumney, petitions

> "for restoration to the office of Providing Flowers and Sweet Herbs for the court, granted by the late King to herself and her late Mother, who with her own two sons was slain in the battle of Naseby."

The petitioner was sworn in "to the place of garnisher and trimmer of the chapel, presence, and privy lodgings, in the room of her Mother, Eliz. Burgess," and we are left imagining the rough seventeenth-century camp where a troubled King, no statesman, but a fastidious gentleman, surrounded by his most needed household retainers including the Provider of Flowers and Sweet Herbs, prepared for the disastrous battle which was to rob him not only of men and arms and power but also of this humble and loyal woman servant who brought, even in the midst of battle and the smell of death, a kind memory of green fields and peaceful gardens. Surely Elizabeth Burgess, slain in the service of her King, June 14, 1645, is entitled to add a sprig of the martyr's palm to the bunch of herbs which was her badge of office.

In *Coronation Cavalcade*, by F. Gordon Roe, we are told that at the elaborate Coronation of James II and Mary of Modena in 1685, "among other ceremonies, that of the herb-women, who strewed flowers before the procession, was resuscitated." Her picture, rather grim and sedate, leads the "strip" engraving of the foot procession in Francis Sandford's contemporary account of the Coronation, where she also receives a pleasant paragraph:

> "Two Bredths of Blew Broad-Cloath was spread all along the middle of the Passage, from the Stone-Steps in the Hall, to the foot of the Steps in the Choir, . . . which Cloth was Strewed with Nine Baskets full of

> Sweet Herbs and Flowers, by Mary Dowle, Strewer of Herbs in Ordinary to His Majesty, assisted by Six Women, two to a Basket, each Basket containing two Bushels."

That William and Mary and, after them, Queen Anne, walked on sweet herbs to their crownings is shown by a manuscript order, dated April 11, 1702 (and quoted by W. J. Passingham in his fine *History of the Coronation*) " for a gowne of scarlet cloth with a badge of Her Majesty and a cypher on it, for the Strewer of Herbs to Her Majesty as was provided at the last Coronation." Queen Anne, who was crowned April 23, 1702, had a regular household Strewer of Herbs, who is listed, in John Chamberlayne's *Magnæ Britanniæ Notitiæ*, 1708, with such special servants as Moletaker, Operator for the Teeth, and Yeoman Arras-Worker.

The next recorded Herb-woman appears on September 22, 1761, in the famous " muddled " Coronation of George III, where everything, after elaborate preparations, seems to have gone wrong, even to the King's Champion's horse, who had been carefully schooled to back out of the King's presence, insisting upon backing *into* Westminster Hall, to the hysterical amusement of many beholders. No details as to the behaviour of the King's Herb-woman have come down to us from that confused day, but we do know that her name was Honor Battiscombe, and that Miss M. Rayner, at a salary of 24 pounds per annum, was Herb-Strewer by appointment to the King's household. This is interesting, for it shows that by now the office of Coronation Herb-Strewer had become purely an honorary one, not to be fulfilled by the humble everyday strewer of herbs of the King's retinue.

So we come to the next, last, and most colourful of King's Herb-women. George IV, while Prince Regent, had as his Physician Extraordinary William Fellowes, M.D., of Bath. This Dr. Fellowes had five sons, among them Commander William Dorset Fellowes, R.N., Secretary to the Lord Great Chamberlain at George IV's Coronation, and a daughter, to whom the Prince Regent is said to have been much attached. And when it was time for that Prince to ascend the throne, and preparations for a magnificent Coronation were in full swing, with court ladies competing for the office of Herb-Strewer in the Royal procession, it was discovered that Miss Fellowes had been promised that honour by the King himself while he was still Prince of Wales. Thus, on July 19, 1821, at about eight o'clock in the morning, Miss Fellowes, followed by her attendant maids, the Misses Garth, Collier, Ramsbottom, Hill, Daniel and Walker, was led into Westminster Hall by her brother, Secretary to the Lord Great Chamberlain, there to await the arrival of the King and the start of the procession to the Abbey. They were welcomed with enthusiastic applause, as the only women to appear in the Queenless procession and as the heralds of a Coronation performance which had been designed to display all the mediæval pageantry in the best early nineteenth-century flamboyant manner. Here are two contemporary glimpses of them (chosen from a wealth of material) as they appeared to a young Quaker girl, Miss Katherine Fry, who watched the procession from a stand by the Abbey, and to the painter, Benjamin Robert Haydon, who sat in Westminster Hall to see the King's Banquet:

> " A second gun fired, the procession had left the hall. . . . In a few seconds the Herb Woman and her Maids were indistinctly seen to the left, coming slowly toward us, with a good deal of stage effect in their action in throwing the flowers (rose leaves). Miss Fellowes herself was dressed in a plain white satin gown, the body fitting tightly to her fine figure; over it a scarlet mantle lined with white, fastened on each shoulder and confined into a peak behind at the bottom of the waist. Its effect was extremely novel and elegant. . . . She held a small gold wire basket lined with satin, from which she scattered flowers, and when it wanted filling turned round and it was replenished by her attendants from their more ample supplies. These attendants were six young girls of our own age, carrying a large basket between two; these also were of gold wire, in shape and size like clothes baskets."

> " After two or three hours waiting [in Westminster Hall], the doors opened, and the flower-girls entered, strewing flowers. The grace of their action, their slow movement, their white dresses, were indescribably touching; their light milky colour contrasted with the dark shadow of the archway, which, though dark, was full of rich crimson dresses that gave the shadow a tone as of deep blood; the shadow again relieved by a peep of the crowd, shining in sunlight beyond the gates. . . ."

A Gothic-novel sort of Coronation George IV's may have been, with an embarrassing richness of waving plumes, gold lace and trailing velvet, and the jovial monarch himself looking, as Miss Fry wrote, "truly kingly, but also too much like a King in a penny picture." But let us, rather than joining unkindly critics in comparing it to the marvels of Astley's Circus, be grateful to that Coronation for presenting, with a swan-song burst of splendour, so many of the ancient customs of a King's crowning which shall undoubtedly be seen no more. And let us thank Miss Fellowes in her scarlet mantle, and the attendant maids in their garlanded white satin, for carrying out the Herb-women's honourable office with a grace and charm which live after them in many contemporary accounts, before consigning their golden baskets to that Limbo which is the more fragrant for their presence.

Bibliography

Calendar of State Papers, Domestic, Charles II, 1660-1.

Francis Sandford, Lancashire Herald of Arms. *History of the Coronation of King James II and Queen Mary.* London, 1687.

Richard Thomson. *A Faithful Account of the Processions and Ceremonies observed in the Coronations of the Kings and Queens of England, exemplified in that of their late most sacred Majesties King George III and Queen Charlotte.* London, 1820.

Arthur Taylor. *The Glory of Regality.* London, 1820.

A Brief Account of the Coronation of His Majesty, George IV. London, 1821.

Katherine Fry. Letters to her sister, July, 1821, in *The Daily Telegraph* [London], Saturday, April 17, 1937.

Sir George Nayler, Garter Principal King at Arms. *The Coronation of King George IV.* London, 1839.

Leopold G. Wickham Legg. *English Coronation Records.* Westminster, 1901.

Notes and Queries, series 10, Vol. XXII, October, 1909.

Eleanour Sinclair Rohde. *Herbs and Herb Gardening.* London, 1936.

J. G. Noppen. *Royal Westminster and the Coronation.* London, n.d.

W. J. Passingham. *A History of the Coronation.* London, n.d.

F. Gordon Roe. *Coronation Cavalcade.* London, n.d.

F. Gordon Roe. *Some Coronation Prints*, in *Apollo* for May, 1937.

WINTERGREEN

MRS. GEOFFREY G. WHITNEY

Gaultheria procumbens, L.

N. O. Ericaceæ
Subfamily III, *Ericoideæ*
Tribe II, *Andromedeæ*

COMMON NAMES: Checkerberry, Teaberry, Aromatic Wintergreen, etc., etc., etc.

THE genus *Gaultheria* was named by Kalm, about the middle of the eighteenth century, for Dr. Hughes Gaultier (the "h" having crept into its title unaccountably in later writings), a naturalist and physician of Quebec.

Among the many descriptions referred to, none seems more fitting than this from Harriet L. Keeler, in *Our Northern Shrubs* (Charles Scribner and Sons, 1922):

> " A small evergreen shrub, with stems creeping on or below the surface, and short, erect flowering branches three to five inches high bearing at their summits a crowded group of aromatic leaves. Found in cool, damp woods from Newfoundland to Manitoba and southward to Michigan and Georgia."

Here follows a botanic description, and then:

> " The leaves of the Wintergreen as they put forth from the top of the upright stem are as near ambrosia as anything our northern climate can produce. The spicy aromatic flavor appears in leaf and stem and fruit, but is most delicate and delicious in the young leaves. One who has sought and eaten them in childhood may wander far from his native home, may forget much of his youth; but the picture of the leafy glade of the forest where he picked the Wintergreen, the carpet of shining leaves, the twin red berries hanging upon thir stems, will never leave him. . . .
>
> " The genus is mountain born and bred; its headquarters are the slopes of the Andes, where nearly one hundred species appear."

Bailey's *Encyclopedia* mentions the genus as having:

> ". . . About 90 species in the warmer and subtropical regions of Asia, Austral., and in Amer. from Canada to Chile. Some have edible fruits, and an aromatic oil used in perfumery and medicine is obtained from G. procumbens and several Asiatic species."

Pursh lists the Creeping Snowberry (*Chiogenes hispidula*) as Gaultheria Serpyllifolia, and also lists its two former names,—*Vaccinium hispidulum* (Willd. sp. pl.) and *Arbutus filiformis* (Lam. encycl.) — an awesome number of aliases to be attributed, as if it were a villain, to this most modest, retiring and untamable of all our evergreen shrublets. The leaves of the Snowberry are said to have the same aromatic flavor as those of Wintergreen.

From Miss Keeler one may turn, with no little amazement, to find our present protagonist appear somewhat in the light of an " elegant " pot or border plant in this from *The Botanical Cabinet: Consisting of Coloured Delineations of Plants from All Countries, with an Account of Each and Directions for Management, etc., etc.*, by Conrad Loddiges and Sons (Vol. IX), — *The Plates by George Cook,* —(1824)— " *Even Solomon in all his glory was not arrayed like one of these.*"

> ". . . It has been known in our gardens for a considerable time and is perfectly hardy. . . . It may either be kept in a small pot or planted in a border in peat earth. It does not exceed four or five inches in height

but spreads a good deal, and continues in flower a long time during the latter part of summer, at which time the ripening berries, produced from the preceding year's blossoms, add much to its beauty. These frequently remain the whole winter, and with the fine dark evergreen leaves, look very pretty in that season. It propagates itself readily by its creeping roots. We are told that in some parts of America the leaves are used as a tolerable substitute for tea."

Pursh, in his *Flora Americæ Septentrionalis* (1848), describes it thus:

> " A little shrubby plant, resembling seedling plants of Kalmia latifolia; flowers white; berries red, eatable, and known by the name of Partridge-berries. The leaves, if properly cured, make a most excellent tea, for which reason it is likewise known by the name of Mountain-Tea."

Even so late as 1853, in the third edition of *Flora Cestrica: an Herborizing Companion for the Young Botanists of Chester County, State of Pennsylvania,* William Darlington, M.D., LL.D., etc., writes of it only as follows, with no serious mention of any medicinal or commercial uses:

> ". . . suffruticose humble evergreens . . . Obs. This spicy little plant — though rare here — is very abundant in New Jersey; and is much used in the preparation of medicated syrups, and popular diet-drinks."

Before that, however, it is shown to have stepped up into some therapeutical esteem by the following from *A Catalogue of the Medicinal Plants, Indigenous and Exotic, Growing in the State of New-York, with a brief account of their Composition and Medical Properties* (1848), by Charles A. Lee, M.D., Professor of Materia Medica in Geneva Medical College and the University of Buffalo, etc.:

> ". . . Berries are edible and have a very pleasant flavor. The oil is recently ascertained by M. Cahours to be a salicylite of the oxide of methyl, and composed of an acid called the salicylic, (hitherto found only in the oil of spirea), united with the ether of wood-spirit (Methylic Ether)."

In more recent years, our little plant has come into its own as to its uses both " For Physic, for Flavor and (according to the note in Bailey's *Encyclopedia*) for Fragrance."

Every one knows it well in candy and cough drops, and in the lowly and vulgar (albeit pleasant) chewing gum; in mouth washes, tooth pastes and powders; and as a disguise for ill-tasting medicines.

It is, however, as a full-fledged and acknowledged " Official " member of the U. S. Pharmacopœia that its greatest pride and glory has been reached — *Oleum Gaultheriæ* (Oil of Wintergreen) being used both externally as a liniment and internally as a remedy for the vague and still undefined " rheumatism " and kindred ills.

Its official title is *Methylis Salicylas* (Methyl Salicylate to the laity), of which it gives about 99% as one of its constituents, — " The remainder consisting of a paraffin called triacontane, an aldehyde, an alcohol and an ester." (Youngken)

It has rivals, however, and must needs share its glory with the following adulterants: "Petroleum products, synthetic methyl salicylate, oil of Sweet Birch and oil of Sassafras." (ibid.)

The oil of Sweet Birch (*Betula lenta*) appears to be its most insidious rival, as it is said actually to steal our hero's name and masquerade deceitfully as Oil of Wintergreen. However, the Sweet Birch is a pleasant tree in itself, and its twigs delightful to munch, so perhaps this " humble evergreen " does not mind being thus traduced.

Mrs. Grieve says, in *A Modern Herbal,* " To the alcohol and ester are due the characteristic odour of the oil." The " part used " is, properly, the leaves only, " harvested in September," but the stems, and sometimes the entire plants are also used in distillation of the oil.

By now we have wandered far afield from the little unassuming ground cover which, unlike most of its near relatives who *call* themselves " evergreens," is actually and definitely so — its only concession to the season being the turning of most of its leaves to a reddish bronze, especially at the edges, as the autumn advances and before winter actually comes upon it.

It does not sulk nor burn, and is astonishingly easy to transplant. Often a bit of it comes along, free and as a present, mingled with the roots of some native shrub got from a dealer in " wild " plants, and one may dig, yes, and tug and wrench it away from the other entwining roots, in the most heartless manner; yet, when re-planted in a kindly situation amongst its brothers, it never " turns a hair ", but settles right down as if it had always been there — as if, in fact, it felt grateful for the change and for having room to stretch its legs unmolested by the stronger roots of its erstwhile tall companion.

Being of the Ericaceæ, it seems scarcely necessary to say that it needs a soil acid, loose and sandy.

It does particularly well under the shade of coniferous trees; and especially is it happy in the shade of hemlocks, where its height is lowest, and its close clustering groups are at their best — so low, so green, so dense, that the taller and larger leafed growth of the same plant found here and there under deciduous trees seems gross and vulgar by comparison.

This does not mean that it wants to be squashed beneath low growing branches of young Hemlocks — nothing likes that — but it does do best of all when the branch of a taller Hemlock waves a foot or more — but not less — above it. Even in the shade of Rhododendron and Laurel, as an underplanting for which it is so often recommended, it does not do itself full justice.

Bibliography

American Berries of Hill and Dale, National Geographic Magazine, February, 1919.

American Medicinal Plants, Charles S. Millspaugh, M.D., 1887.

Gray's New Manual of Botany, Robinson and Fernald, 1908.

Materia Medica and Pharmacology, David M. R. Culbreth, Ph.G., M.D., 1927.

Wild Flowers of the Alleghenies, Joseph E. Harned, 1931.

And a host of other modern Wild Flower books.

Books referred to in the article:

Bailey's Encyclopedia of Horticulture, 1922.

Botanical Cabinet, The, Conrad Loddiges and Sons, Vol. IX, 1824.

Flora Americanæ Septentrionalis, Frederick Pursh, 1814.

Flora Cestrica, William Darlington, M.D., LL.D., 3rd Edition, 1853.

Hortus, L. H. and E. Z. Bailey, 1930.

Manual of the Botany of North America, A, Laurence Johnson, A.M., M.D., 1884 (Wood's Library of Botany).

Medicinal Plants, Charles A. Lee, M.D., 1848.

Modern Herbal, A, Mrs. M. Grieve and Mrs. C. F. Leyel, 1931.

Our Northern Shrubs, Harriet L. Keeler, 1922.

Text Book of Pharmacognosy, A, Heber W. Youngken, A.M., Phn.M., Ph.D., Sc.D., 1936.

Old English Herbs from the Olitory at Stonehurst, Shanty Bay, Ontario, Canada.

Balm

Basil

Fennel

Marjoram

Mint

Sage

Parsley

Sage

Summer Savory

Tarragon (Leaves)

" (Ground)

Thyme

The Herbs furnished from the Olitory at Stonehurst, may be used dry at the table to flavour soups, sprinkle potatoes, etc., the same as you use pepper and salt.

These herbs are grown in suitable soil, in sunshine or shade as is most fitting, harvested at the right time, dried in the proper way, carefully winnowed and finely powdered. They are absolutely pure and much stronger flavoured than herbs usually obtainable.

---

1 Jar..................	$ .50
Box of 6 jars.....	2.50

These herbs make an attractive and useful Christmas present. A "Herbal" of historical value and containing many receipts from the old herbals, beginning with "Dioscorides" of the first century A.D. up to the XVIII century, is sent with each box.

Toronto Office

THE GREENMAN

184 Bay Street Elgin 2953

The Greenman

Printed in Canada

The HERBARIST

A Publication of

The Herb Society of America

No. 5

For Use and for Delight

BOSTON, MASSACHUSETTS

1939

The Catnip Man of Boston.

The Herbarist for 1939
A Limited Edition
Written and Published by
The Herb Society of America
Price .$1.00

Copies of the 1935, 1936, 1937 and 1938
Herbarists may be purchased
at $1.00 each.

Address
The Herb Society of America
Horticultural Hall
300 Massachusetts Avenue
Boston, Massachusetts

CONTENTS

Dyeing Skeins of Wool

woodcut from:

Plictho de l'Arte Tentori by Gianventura Rosetti, Venice, 1548

In which Indigo, then a rare dye imported from America, is discussed

INDIGO, THE TRUE BLUE

MARTHA GENUNG STEARNS

PERHAPS you will remember the little villages in Italy perched up on impossible rocks or clinging to a mountain-side, which have been sun-baked for so many centuries that they have mellowed to terra-cotta and merged with the sun-scorched rock. If you have climbed up to one of these primitive little places, you have come through a crumbling arched gateway into the village square, and there, up in the campanile of the church, you see a bit of intense blue, repeating the higher blue of the sky. It is the face of the clock, an ancient timekeeper, sometimes with but eight numerals, so that in the twenty-four hours its one hand must make the circuit three times. It marks only the hours, perhaps because minutes don't matter very much when you live above the world. The face is painted blue, the color of eternity, of infinity, to remind one who looks at it that time will one day cease to be.

It seems as if the most loved and familiar things in nature are one sort of blue or another; the distant blue of mountains and cloud shadows and the sea, and the same hue comes into our immediate foreground in many a bird and tiny flower, even to the little blue eggs of robin and hedge-sparrow. In all the handicrafts, the work of a man's hands is intimately bound up with nature, as he takes materials which have grown out of the earth and brings them by simple processes to a useful and beautiful end. All through the history of the arts and crafts, we find indigo as a staple pigment and dye from the earliest days down to the present time; there is no other natural blue so reliable or running so true.

In Christian symbolism, which is really a picture language of astonishing range and richness, every color and almost every natural object has its meaning. Blue is not only the color of Our Lady's garment in the works of the old painters: it is the soul's color, the color of faith. The flower of nigella, or love-in-a-mist, was a symbol of the soul, a blue blossom enclosed in a feathery green sheath, the veil of the mortal body. In the world's universal language green is the earth color and blue the symbol of heaven.

Blue, as we think of it in the arts, is almost synonymous with indigo. In the Middle Ages, there were three great natural sources of blue. One was mineral, a copper ore called Azurite, or lapis lazuli, which had to be reduced to powder and mixed with some vehicle which would carry the particles; and there were two plant sources, Woad, always considered as secondary, and true Indigo, which stood severe tests and was the colorist's standby. There was not much difference, chemically, between woad blue and indigo blue, and the woad which grew readily in northern countries was substituted for the Oriental indigo for domestic purposes. This was partly due to the fact that the ashes of woad, when the stems and waste were burned, were the best source of potash and lye; and as such, woad has been used down to the present day as an adjunct to dyeing with indigo in potash solution, so that the terms are sometimes interchanged and consequently puzzling. But woad was a disagreeable plant, exhausting the soil so that new tracts must constantly be given up to it, and also fermenting in the dyeing process and becoming extremely odorous. Daniel Thompson says, in "Materials of Medieval Painting,"

> "The other great blue that the Middle Ages inherited from antiquity was indigo, a colouring-matter known to the ancient Egyptians and imported by them, and later by the Greeks and Romans, from India (whence its name). It was extracted from certain plants native to India known to modern botany as *Indigoferae,* the bearers of indigo. It was used primarily as a dye, but as a by-product of dyeing with it a blue pigment was formed, the colour *indicus* or *indicum,* which became in Italian *endeco* or *endego,* and in English 'Indigo.' The importation of indigo from the Orient continued through the Middle Ages, and the best qualities in the European market were known as Bagdad indigo or Gulf indigo."

Indigo by itself is a dark purplish color and must be thinned and lightened to get a true, clear blue. Most of the indigo in the market still comes from its ancient home, India. The plant *indigofera sumatrana* is grown from seed sown about the end of April; by the

Design by Elizabeth Hartwell of Edgartown, Mass. 1786
In Three Shades of Blue

middle of June it is from 3 to 5 feet high, and the first crop is cut then, another being ready by August in that climate. It is cut early in the morning, taken to the factory and steeped in water from 9 to 14 hours, when the liquid is run off into another vat, where it is violently beaten; the object of this is to bring as much air into it as possible, and in the hand-process this whipping was done with bamboo sticks. The precipitated indigo is allowed to settle like a sediment or mud to the bottom, and the water run off. After being strained and sterilized by boiling, it is dried and cut into blocks. The color comes from the leaves alone, the stems yielding practically none. In its dried cake form it is easily handled and exported, and thus it travelled over the world, coming to America in the days of the earliest settlements.

India brought dyeing to a fine art, and was the first Eastern country to export dyes into Europe. Italy was soon learning her secrets and was noted during the Ages of the Faith, when the Church was the great patron of the arts, for her magnificent fabrics, especially Florence, which was the city of master-dyers. By the slow progress of time and commerce, the art of textile weaving and dyeing moved northward by way of Lyons and Arras, Bruges and Spitalfields, and what would these fabrics have been without their glorious blues? Indigo was preëminently the dyer's color; in illuminated books and on canvas and especially for enamels, ultramarine was more desirable, a powder of ground lapis lazuli from Persia which went through long processes to reach the painter's brush in workable form. An Italian book on dyeing, printed in Venice in 1548, spoke of "endego" as a rare dye from the Americas, and the renowned traveller Marco Polo speaks of it.

When America entered the company of nations, indigo was well known, and the simple but laborious hand-processes went on for more than two centuries. What we call by the covering term of "Puritan gray" is said to have been a wool fabric woven by the formula of one part good white sheep's wool, one part black sheep's wool, and one part of "taglocks" and other discolored bits spun and dyed blue in the indigo pot, all three woven into a dull mixture of practical worth but little beauty. A blue dye-pot was said to see as much service as a teapot; and these old earthenware dye-pots with a blue sediment in the bottom only needed more water to be boiled up and used for many household needs, sometimes sweetened by flag-root to counteract the strong odor. In an austere society which frowned upon bright colors and display, the humble blue was allowed, and this accounts for the prevalence of the honest, hold-fast color in early needlework, woven bed-covers and garments. The indigo peddler appeared once or twice a year, bringing the dye in lumps, and the best quality cracked apart and showed a glint of gold. It was tied up in a cloth and pulverized by pounding, and a little went a long way. The depth of color depended on how long the fabric remained in the pot, or the number of dippings.

Eliza Pinckney, the energetic mistress of Belmont, near Charleston, South Carolina, demonstrated in 1744 that she could grow indigo successfully, and a minute description of her method is given in her life by Harriott Horry Ravenal. She obtained seed from the East Indies and devoted a whole crop to raising her own seed, much of which she gave away; she writes: "Out of a small patch of indigo . . . besides saving a quantity of seed, made us 17 pounds of very good indigo," and in 1753 the neighboring planters formed themselves into the "Winyah Indigo Society," which flourished for years and exported enormous quantities to England, receiving a bounty from the Crown.

A less successful experiment was conducted at the Pine Mountain settlement school in Kentucky in 1925, with seed from France. The old hand-process was carefully adhered to, a pound of leaves to a gallon of water; but the yield in dry weight of cake indigo from 12 pounds of leaves was just short of one ounce. Considering that Madras indigo can be bought for $2 per pound, it would not seem practical to spend the time, labor and expense necessary to extract indigo; but the fact has been established that it is entirely possible to get excellent indigo from American-grown plants and American seed.

Dyeing methods were revolutionized by the discovery in 1834 of the aniline process by a German chemist who noticed that a bright blue color resulted from the action of a bleaching powder on a distillation of coal-tar. Artificial indigo was worked out in 1878; and now the aniline process has displaced the old vegetable dyes, giving a vast range of shades; but with the passing of the old-time method, we have lost another contact with nature and her own soft and subtle shades which had so much charm.

A NOTE ON THE HORSE CHESTNUT

HEBER W. YOUNGKEN

THE common Horse Chestnut (*Æsculus Hippocastanum* L.) which has been long admired as an ornamental shade tree and whose seeds contain an abundance of starch which could be used as food for animals and human beings, were economical means of depriving them of the bitter principle devised, yields a valuable glucoside which is being utilized at present in sun-tan preparations. This glucoside, called "æsculin," which has been reported to occur in all parts of the plant, is claimed to be found most abundantly in the bark and fruit wall. It was discovered by Minor in 1831. Æsculin is very sparingly soluble in water, forming a faintly blue, fluorescing solution. It is now being used in a 4% solution or ointment to protect the skin from sunburn. It has the property of absorbing ultra-violet rays, which it then gives off gradually. In addition to its use as a prophylactic against sunburn, it is also used in heliotherapy and as a prophylactic against snow blindness.

WOAD

ELIZABETH WADE WHITE

BY VIRTUE of a vigorous yellow-blossomed plant the Metropolitan police of England, in their uniforms of dark woad-vat blue, are the inheritors and descendants of that shaggy, shadowy race which the Romans found, some years before the birth of Christ, inhabiting the island of Britain. It is said by some authorities that the word "Britanni" is the latinization of the Welsh "Brython" (from bryth= paint), and thus means "the painted men," those of whom Cæsar wrote:

"All the Britains, indeed, dye themselves with woad, which occasions a bluish colour, and thereby have a more terrible appearance in fight."

The herb woad *(Isatis tinctoria)* probably holds a more important place in English economic history than any other single plant (excepting, of course, wheat), having for almost two thousand years yielded from its dark green juicy leaves the blue pigment that progressed logically and expandingly from the staining of savage English skins to the dyeing of English wool and cloth.

Little enough is known about the preparation of woad-dye by the early Britons, except for two interesting facts: one, that the plant seems not to be a native of the British Isles, and must therefore at some remote time have been imported from the continent and established in cultivation; the other, that not only blue, but also black and green colouring were obtained from woad in some primitive fashion by the pre-Roman Britons. Pliny says:

"In Gaule there groweth an herb like unto Plantain, and they call it Glastum (*i.e.* Woad), with the iuyce whereof the women of Britain, as wel the maried wives, as yong maidens their daughters, anoint and dy their bodies all over, resembling by that tincture the colour of Moores and Ethyopians: in which manner they use at some solemne feasts and sacrifices to go all naked."[1]

[1] Pliny, *Natural History,* Book 22, Chapter I.

And Ovid[2] pleasantly calls our earliest ancestors "virides Britannos" (the green Britons). Their ways of making these colours were probably, for black, simply rubbing themselves with the juice of the leaves, and for blue and green, adding wood-ashes or quick-lime, respectively, to an infusion of the young leaves in hot water.

Undoubtedly, some of the cloaks and togas in which the Roman invaders stood to watch the religious antics of the paint-clad Britons were dyed with the selfsame pigment, for woad is a plant that has flourished in most of the temperate regions of the world and has been used by man since the beginning of botanical history. Hippocrates, the great Greek physician, writing about 460 B. C., speaks of the curative value of woad in the treatment of ulcers, and the later classical writers, Pliny and Galen, mention it as a dye when they detail its medical properties. Whether woad or its only rival, indigo, provided the earliest Egyptians with their blue pigments is a question, but we know from the Oxyrynchus papyri, of the first century of the Christian era, that woad was cultivated as a dye-plant in Egypt at that time.

The recorded history of woad in the economics of England begins with the beginning of the thirteenth century, and continues, hand in hand as it were, with the history of the most important of English commercial staples, wool. It is also linked closely with the development of the dye-and-cloth trade on the continent.

From the start of the woad industry, English fields could not produce enough to supply the dyers' needs, and much woad was imported from France and Italy, from the Duchy of Jülich and from Thüringia in Germany, the last the richest woad-growing district of the world. In 1213 the duty paid on woad entering English ports amounted to nearly £600. Later in the century, during the reign of Henry III, one of the first "Buy British" experiments was tried: Parliament ordered that only English-made woollen clothes should be worn in England. This caused bitterness abroad and a consequent interruption of the woad trade, but not much time elapsed before the superiority of the Flemish over the English weavers was acknowledged, and toward the middle of the fourteenth century, in Edward

[2] Ovid, *Amorum*, Book II, xvi, 39.

III's time, another experiment for the health of the English cloth industry was made, which consisted of the introduction into England of numbers of Flemish weavers and woad dyers. It is interesting to note here, in order to obtain some idea of the international importance of the dye trade of this time, that in Florence, in 1342, woad played its part in the formation of the earliest political union of oppressed workers. This was the company of dyers, associated with the powerful Wool Guild in the capacity of an unenfranchised branch, which, led by the woad-dyers, its foremost members, carried on a dramatic series of attempts to assert its rights as an independent organization. The dyers finally succeeded in establishing their own Guild, but their triumph was short-lived, for the woad trade and its associated activities were too valuable to the merchant princes to be allowed any freedom along the lines of self-government. Indeed, the production, sale and use of woad was hedged round with strict regulations in every country which was so fortunate as to possess this valuable plant as one of its natural resources. In Germany woad went, according to law, from the farmer who grew and harvested it to the refiner who prepared it for use, and so to the woad merchant, a licensed and privileged person who distributed the woad for export or national use, and each of these changes of hands was attended by a whole series of taxes, duties, and restrictions. In France, which produced a particularly fine quality of woad from the provinces of the Somme, Languedoc and Lower Normandy, the woad industry was so highly organized that each town in the woad districts had its own set of regulations governing the trade, and national ordinances controlled the export of woad and required that all dyed fabrics should be stamped with labels indicating the dyes used and the name and town of the dyer. An interesting piece of sculpture at Amiens Cathedral shows two mediæval woad merchants exhibiting their sack of woad-balls, and it was these balls, or *cocaignes,* into which the crushed leaves of woad were formed after harvesting, that gave to the Languedoc its name of *pays de cocaigne,* which has since come to signify any land of great richness and productivity.

In England there were many city and national ordinances governing the traffic in woad, and officials such as brokers, measurers, and price assayors were appointed to maintain the legalities of the trade. Hakluyt's *Libelle of Englyshe Polycye*, a poetic review of English commerce published in 1436, has the following reference to the woad trade between Genoa and England:

"The Janueys comyne in sondry wyses
Into this londe wyth dyverse marchaundyses
In grete Karrekkis arrayde wythouten lake
Wyth clothes of golde; silke and pepir blacke
They bringe wyth hem and of woad grete plante."

In 1567 another great influx of foreign weavers, this time Dutch and Walloon refugees from the Spanish government of the Netherlands, occurred in England. So in the time of Elizabeth the English cloth-trade reached great heights and with it the woad industry, large quantities of the dye being produced in England, especially in the counties of Somerset and Lincolnshire and parts of Cambridge. In this its golden era, woad was so valuable a crop that its cultivation was strictly regulated by royal proclamations, a register of all woad farmers being kept, and each one of these farmers being obliged to pay a tax of twenty shillings a year to the queen. Elizabeth detested the ammoniacal odour which issued strongly from woad during its process of fermentation, and so it was decreed that no woad should be grown within five miles of any of her residences or of any city, market town or main highway. But woad-growing was otherwise encouraged by a government which appreciated the fact that forty acres of this valuable plant were sufficient to keep one hundred and forty persons constantly employed.

At about this time, when the world's horizons of trade and knowledge were so rapidly expanding, the woad industry felt the first thrust of what was eventually to cause its decline and practical disappearance. This came in the form of "blue ynde" or exotic indigo, which was brought from the Orient by way of the Mediterranean countries, and which seemed to give a very brilliant blue dye at a much lower cost than woad. But many were suspicious of this new thing, particularly since its colour did not have the enduring qualities of the woad-blue, and woad continued to be cultivated in England. It is said that in the seventeenth century the produce of one acre of woad sold for from £6 to £20 a ton, and so much was grown in England at this time that its importation from France seems to have ceased altogether after 1638. Toward the end of the eighteenth century a very large woad farm, having more than 1,100 acres under cultivation and an eight-year system of crop rotation, was established at Brothertoft near Boston in Lincolnshire. There were other important woad farms in this same district, founded at about this date or later, one at Parson Drove near Wisbech in Cambridgeshire which survived until 1914. The method of preparing woad for the dyer which these farms employed was hardly different, except for efficiency and range of equipment, from the mediæval process. The plants, which are biennial or perennial, were allowed to grow about seven inches high, in rich, well-drained soil, then the first crop of leaves—there were usually two and sometimes three crops a year—was picked, by hand, and conveyed to the woad mill. Here, in the first department, the roller house, the leaves were crushed to a pulp by large drumlike wheels revolving in a stone-paved circular depression around a central axis. These wheels were generally pulled round by a horse. After this the wet pulp was heaped up to drain for a short time, and then, by the always black hands of the woadmen, was formed into balls four or five inches in diameter. Wickerwork trays or "fleaks" were filled with these balls and conveyed, on the padded heads of other woadmen, to the drying ranges, which were rough, outdoor wooden frames, roofed over but open on the sides. Here the woad-balls were left to dry for from one to four weeks. Then they went back to the roller house and were ground to powder under the same wheels that had pulped the green leaves, and from there the pulverized mass was taken to the couching house, spread on the stone floor in a layer two or three feet deep, sprinkled with water and left to ferment for about nine weeks. This fermentation process was at first a furiously bubbling and steaming one, sending up the "noysome savour" which was so distasteful to Queen Elizabeth, but gradually, and helped on by frequent stirring and turning over, the mass settled down into a dark-coloured clay-like substance, when it was ready to be broken and sifted into powder, packed in large wooden barrels, and sent off to the dyers.

Woad was the universal dye of the Middle Ages, for it alone, before the coming of indigo, supplied the range of blues which were in constant demand, and also in combination with other dyes it produced other important colours. Woaded fabrics were boiled with alum or tartar and then dyed with yellow, to make a fine green, or with cochineal or madder for various shades of red, purple or violet. And woad-black, obtained by the use of madder, logwood or nut galls, was considered so superior to other blacks that it was in some instances legally protected from imitation or false representation. The mediæval woad vat was prepared by heating woad and its mordant, usually alum or potash, with water and keeping the mixture at a high temperature for several hours. During this heating process a blue scum rose to the surface, and this was collected, dried and powdered and, known as "flower of woad," was used as a blue paint by the artists and illuminators of the Middle Ages. Cloth to be dyed was immersed in the woad vat for a period varying in length according to the depth of colour desired. Azure, plunket, and watchet were the lighter shades of blue, and after longer immersion came the fine dark blue which is always associated with woad. That dark blue still comes from the woad vat to colour the uniforms of England's police force, but, strangely enough, the large quantity of woad used in the modern vat has no longer anything to do with the dye which it produces. For indigo has finally and overwhelmingly taken the place of woad as a colour agent in the obtaining of blue dyes, and woad is used in the vat simply because of its chemical ability to reduce the insoluble blue indigotine to a soluble substance known as indigo white. Woollen materials immersed in this substance come out slightly yellow in colour and, after they have been exposed to the air for a while, take on the required shade of blue. It is sad that so venerable and valuable a plant as woad should be degraded to this position of anonymous, if useful, servitude to an outlandish rival. But, however definitely indigo has usurped the place of woad in commercial usage, still woad is an easy and rewarding plant to grow, and, once established in the garden, can be used for small-scale home dyeing of woollen fabric without undue complications. Dr. Jamieson B. Hurry, in his inclusive study, *The Woad Plant and Its Dye,* suggests the following method: Pour one gallon of nearly boiling water onto half a pound of woad leaves, cover the vessel closely and keep it at a temperature of from 100° to 140° Fahrenheit for from ten to twelve hours. In this length of time small bubbles will begin to appear on the surface of the liquid, and now 6 or 7 grammes of recently slaked lime should be stirred in, which will cause an ammoniacal odour. If a piece of woollen cloth is immersed in this mixture for an hour, it will turn pale blue on exposure to the air. After a few hours the addition of a gramme or so of bran will again set the fermentation going, which in its turn can be controlled by the addition of more lime. In this way the dyeing process may be kept in operation for several days. The longer and more often the materials to be dyed are immersed, the darker they grow. The first specimens will be pale blue, the later ones will be dark blue, indeed, almost black.

So the modern herb-growing householder can use one of the oldest dyes known to man, and in doing so, can revive the echo of an industry which for hundreds of years benefited many hundreds of the world's agricultural and industrial workers, and brought pleasure to all who wore or used its products, the woollen fabrics coloured so enduringly by the handsome and seemly plant, woad.

The Herb Society of America has excellent specimens of Woad in its Herbarium.

COLCHICINE
AND THE PRODUCTION OF POLYPLOID ESSENTIAL OIL PLANTS

MABEL L. RUTTLE

THE alkaloid colchicine which is now being used for the production of new varieties of plants is the poisonous principle found in the seeds and bulbs of the autumn crocus *Colchicum autumnale.* The genus *Colchicum* consists of about 30 species which grow in moist meadows throughout Europe, West Asia and North Africa. The name Colchicum comes from the land of Colchis where according to Dioscorides *Colchicum autumnale* grew in large quantities.

Colchicum autumnale blooms in the fall from August to October and brightens meadows with its lilac-colored flowers. The leaves appear the following spring, and the seed pods ripen in May or June. The seeds when fresh are a white color and somewhat sticky, but after careful drying in a dark, not too warm place they take on a dark red brown color. The bulbs are collected in autumn just before blooming.

In meadows where it grows, *Colchicum autumnale* is usually avoided by cattle and sheep, but if by chance a stray bulb or seed pod is eaten, the result is serious illness and perhaps death. The extracted drug, colchicine amorphous, is so deadly that 3 milligrams is sufficient to kill a mouse. The Greeks called the plant "Ephemeron," because any one who ate the bulbs must surely die the same day.

From seventeen hundred on colchicine has been mentioned in medical literature as a treatment for gout. In 1934 Dustin and Lits in Belgium found that in animal cells it arrested cell division, so also did Ludford in England in 1936. In the spring of 1937, Nebel, following these leads and those of others, found that when a stamen hair of *Tradescantia* was immersed in a dilute solution of the drug, the chromosomes in the nucleus could divide but not separate. As a result a new nucleus was formed with twice the number of chromosomes in the normal untreated nucleus. If such nuclei could propagate themselves, a tetraploid plant would result with twice the hereditary units in the treated plant, a goal much sought. Immediately young seedlings were treated with the drug, many were killed, others though temporarily stunted survived, outgrew the stunting and by fall there were growing several tetraploid marigolds, snapdragons, pinks and tomatoes (Nebel, 1937, and Nebel and Ruttle, 1938, *A*). This work has been carried on this summer (Nebel and Ruttle, 1938, *B* and *C*, and Ruttle and Nebel, 1938).

At the same time and independently, Blakeslee (1937) and Blakeslee and Avery (1937), following leads of the Belgian workers and others, treated seeds and seedlings of *Datura, Portulaca, Cosmos,* and many other plants with the drug and recognized that tetraploids were produced just prior to Nebel. Thus colchicine from a treatment for gout has become a drug which may profoundly affect the economy of nations.

To explain the action and use of colchicine, it will be necessary to explain something about chromosomes and heredity, because colchicine reacts primarily with the mechanism of chromosome separation, a more detailed explanation of which has been given by Nebel and Ruttle in 1938, C.

Chromosomes are microscopic cylindrical bodies of protoplasm which carry the bulk of the hereditary traits called genes from one cell to the next and from one generation to another. In figure 1 the nucleus *c* is shown as a ball in the center of the cell *a*. The six chromosomes *d* within the nucleus are more elongated and less condensed than in dividing cells such as that shown in figure 2. In the dividing cell, figure 2, the membrane of the nucleus (figure 1, *e*) has disappeared, the chromosomes have shortened and condensed and appear as six sharply defined bodies arranged in a plate.[1] Each of the chromosomes is split longitudinally. The longitudinal halves now move apart (figure 3), and as individual chromosomes form two daughter nuclei with six chromosomes in each. At all stages of the nuclear cycle, both the resting stage and the dividing stage, the linear order of the genes is maintained so that the order is the same for all cell generations arising from the division of the body cells under normal circumstances.

[1] Plants of different species and genera may have the same or different chromosome numbers.

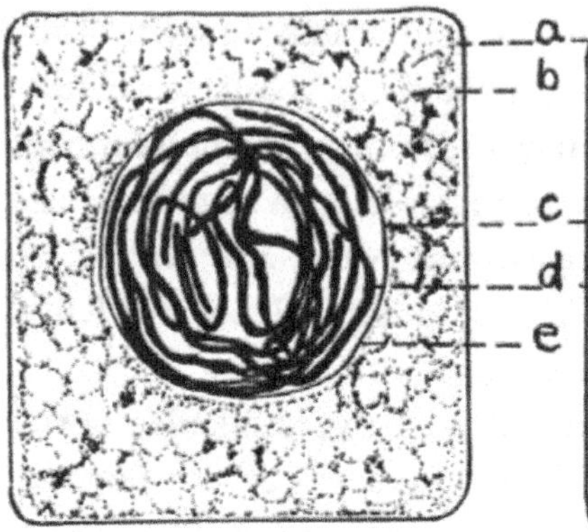

1

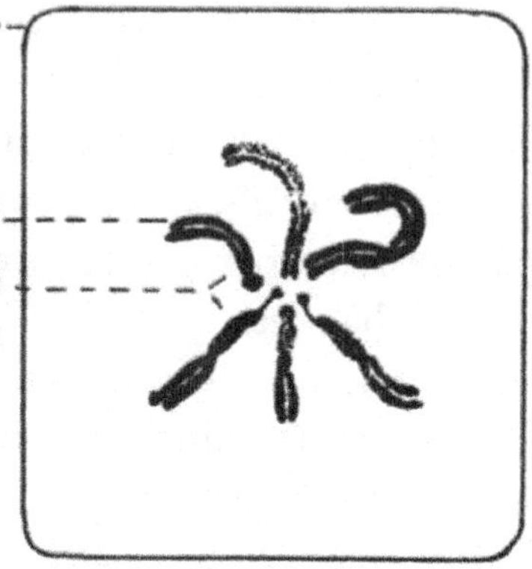

2

3

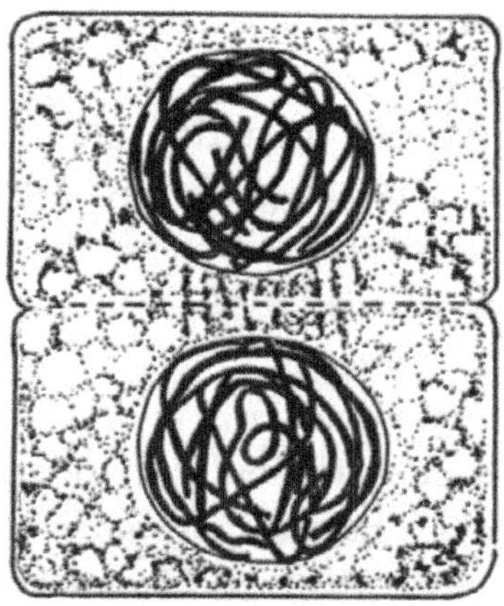

4

5

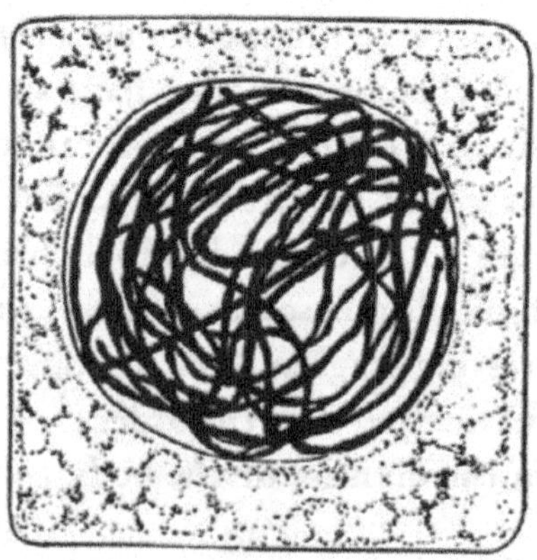

6

Figure Explanations

Figure 1. Body cell of a diploid plant with six chromosomes. The cell is in the resting stage. The chromosomes are so elongated that they cannot be counted. They are clearly split for the next division. *a*, cell; *b*, cell content; *c*, nucleus; *d*, chromosomes; *e*, nuclear membrane.

Figure 2. Body cell of a diploid plant showing the nucleus during the dividing stage. The nuclear membrane has disappeared. The six chromosomes have shortened and thickened and the two parts are preparing to separate. The black set of three chromosomes may be regarded as having come from the male parent, the dotted set of three chromosomes from the female parent. A plant containing one set of chromosomes from the male and one set of chromosomes from the female parent is called diploid.

Figure 3. The chromosome parts of figure 2 have separated and are passing to opposite ends of the cell where they will form new nuclei. Each chromosome part is now a new chromosome and is already split for the next division.

Figure 4. Division of nucleus and cell is completed. Two new cells are formed each with a nucleus which is the exact genetic duplication of the parent nucleus.

Figure 5. If the dividing cell in figure 2 is treated with colchicine, the chromosome parts separate but cannot pass to opposite ends of the cell as in figure 3. Thus a new cell is formed with 12 chromosomes, 6 of which are an exact genetic duplication of the original 6 chromosomes.

Figure 6. The 12 chromosomes enter the resting stage and if there is no further treatment continue growth and division in a normal fashion. A plant or part of a plant made up of cells containing the doubled set of chromosomes is called tetraploid. Such plants are frequently produced by treatment with colchicine.

When a body cell divides (figure 3), the chromosome halves pass to opposite ends of the cell, form a new nucleus, and a separating wall is formed between the two nuclei (figure 4). The separated chromosome halves are not halves in the sense of half an apple, but are complete hereditary units (new chromosomes) produced in the nucleus during the process of cell growth and division. No body cell arises save by the division of another cell. This process, so fundamentally stable, tends to maintain uniform the hereditary characters of an individual.

Changes in the body cells occasionally do occur under natural conditions. Such changes, either in chromosome number or genes in the body cell, when recognized, are called "bud sports." Bud sports, which affect flower or fruit color, have often been selected, but there may be variation in such qualities as time of ripening, for example, the Fisher Peach. Colchicine vastly increases the number of bud sports in the direction of producing sports with higher chromosome numbers, usually a multiple of the number in the parent plant.

Colchicine is a specific and reversible poison for chromosome separation. A normal plant (diploid) with, let us suppose, six chromosomes (figure 2) is treated with colchicine. The chromosomes, in the different dividing cells in the growing point, divide longitudinally in normal fashion, but cannot pass to opposite ends of the cell as in the normal untreated cell (figure 3). Hence, as a result of division, instead of two nuclei being formed, each with six chromosomes (figures 3 and 4), there is formed a new large nucleus with twelve chromosomes (figures 5 and 6). Such a nucleus is called tetraploid, since it has twice the number of chromosomes found in the nucleus of the original parent plant. Tetraploid cells are in many plants capable of surviving and producing one or more branches or parts of single branches which are tetraploid. Seed from flowers borne on tetraploid branches produces tetraploid plants, so that a new type of plant is immediately established. In this way a new tetraploid variety of Guinea Gold marigold has been produced which bears flowers of finer quality than the original diploid, but blooms somewhat later.

The characters of the new tetraploid plants are exceedingly uniform, since all the characters of the parent have been duplicated. Irregularities, however, may occur as in diploids during the production of the egg cells and pollen grains, which may lead to the origin of plants with different chromosome numbers, usually with one or two more or less than the new number. In diploids, plants with one or two more or less chromosomes than the parent are called aneuploids; in tetraploids plants with similar variation are called tetraploid aneuploids. Crossing may occur between diploids and tetraploids which may give rise to triploids and triploid aneuploids. Hence, while the characters of the new tetraploid variety are exceedingly uniform, the possibilities of variation during reproduction are increased rather than decreased.[1] Thus we have obtained triploid and triploid aneuploid marigolds, the progenies from which considerably increase the range of selection.

[1] Reproductive cell division is simply described by Nebel and Ruttle, 1938. *C*, and has been omitted here.

A FIG. 7 B

A. Diploid basil (*Ocimum basilicum*).

B. Tetraploid basil produced by treating diploid seedlings with .16 per cent colchicine for three hours. Note the larger leaves and the heavier stems.

The tetraploid basil (*Ocimum basilicum*) in figure 7, right, was produced by treatment of a normal diploid seedling with .16 per cent colchicine for 3 hours. Figure 7, left, shows the untreated diploid. The stems of the tetraploid are heavier than those of the diploid, the leaves are larger, thicker and somewhat ruffled, the flowers are slightly larger with more fimbriate corolla margins, the pollen grains are larger. Seed sets, but not in such quantity as in the diploids. The odour appears quite different both in quality and quantity. The seeds upon germination produce tetraploid seedlings.

The diploid mint shown in figure 8, right, is a hybrid between *Mentha aquatica* with 48 pair of chromosomes and *Mentha rotundifolia* with 12 pair of chromosomes. It has been grown at the New York Agricultural Experiment Station since 1932, and has always been completely sterile. In the fall of 1937 runners and shoots were treated with .2 per cent colchicine for 15 hours. Numerous tetraploids (amphidiploids) arose from the treated material. There was some variation in the clones, possibly due to the different proportions of diploid and tetraploid tissue.

The diploid (figure 8, right) and tetraploid (figure 8, left) are simi-

B FIG. 8 A

B. *Mentha aquatica* × *Mentha rotundifolia* tetraploid produced by treatment of runners and shoots with .2 per cent colchicine for 15 hours. This plant is fertile. Note the stamens and the larger size of the flowers. The chromosome number is 2 × 60 = 120.

A. *Mentha aquatica* × *Mentha rotundifolia*. This hybrid has been grown at the Station since 1932 and has always been completely sterile. Note that only the pistil shows, the stamens having degenerated early. The chromosome number is 60, 48 from the *aquatica* parent and 12 from the *rotundifolia* parent.

lar in habit of growth, but the stems of the tetraploid are heavier, the leaves larger, the flower spike larger. In the diploid hybrid (with 60 chromosomes, 48 + 12) the anthers always degenerate early; in the tetraploid (with 120 chromosomes, 2 × 48 + 2 × 12) the anthers develop and produce good pollen. Seed set in quantity and tetraploid seedlings are now growing. The odour of the diploid hybrid is weak and cannot be classified as " minty " (Ruttle, 1938 *Herbarist*), while that of the tetraploid is much more pronounced and is more " minty."

Runners and shoots of commercial spearmint were also treated with .2 per cent colchicine for 3 hours. One tetraploid clone was obtained. The clone used for treatment has been grown at the station since 1929 and has always been completely sterile. In the diploid clone the anthers degenerate early, while in the tetraploid anthers develop which contain good pollen, and seed sets. The odour of the tetraploid appears less pronounced and of different quality than that of the diploid. In both these cases the diploid hybrids were completely sterile; by doubling up the chromosome number new fertile species have been obtained.

The marigolds of the Guinea Gold variety originally treated had a distinct but not unpleasant odour. The tetraploids obtained from them do not appear to differ in odour from the parent diploids. The octoploids obtained by treatment of the tetraploids also appear unchanged in odour. Similarly the hexaploid obtained from seedlings of the natural tetraploid French Dwarf which was treated appears similar in odour to the tetraploid sister seedling. In the seedlings from the triploid and triploid aneuploid marigolds, however, which have distinct differences in chromosome number and in genetic constitution, even if of the same chromosome number, there are marked differences in odour.

Thus it appears that tetraploidy may in some genera alter both the quality and quantity of oil. While often the alteration in quality may be disadvantageous, it is possible that in some cases both quality and quantity may be improved. The ultimate criterion of the value of tetraploid essential oil plants depends upon an analysis of their oil when and if sufficient tetraploid clonal material can be grown.

Courtesy of Museum of Fine Arts, Boston

Anthemion — From Attic Grave Stele
Greek 4th Century

Acanthus Mollis Latifolia (Commercially Listed as Acanthus Lusitania). Two-Year-Old Plant Grown from Seed by Miss Margaret Norton, Annisquam, Mass.

SNUFF

ROSETTA E. CLARKSON

"Marierium bruised or rubbed betwixt the hands, and put into the nose thrills . . . draweth downe humors from the head, mundifieth the brayne, causeth to snese and is verie good for them that have left their smelling."

Dodoens: New Herbal
(Lyte's translation, 1619)

THROUGHOUT the centuries, people have drawn "downe humors from the head" by the use of herbs as smoking mixtures or as snuff. The origin and purpose of snuff was to clear the brain. In earliest days no mixtures were made; the leaves and flowers were dried and powdered, or sometimes the juice was expressed from the plant. Some preferred one herb, some another, — the dried leaves of asarabacca (*Asarum Europœum*), of basil, of sweet marjoram, of Syrian mastic thyme, of betony (an ingredient of "Rowley's British Herb Snuff," once very famous), the juice of beets, of rue, ground ivy, the flowers of lily-of-the-valley. In later times, combinations of these head herbs were made, perhaps on the assumption that if sneezing was to be the result desired, four different herbs producing that effect would be four times as effective as one. Neither skill nor art was needed for compounding these mixtures. The commonly recognized head herbs were assembled, according to varied formulæ, and stored in phials, "A half ounce each of sweet marjoram, marum Syriacum leaves and lavender, and a drachm of Asarabacca." Another snuff — "one-half ounce each of leaves of betony and marjoram, and a drachm of Asarabacca leaves."[1]

The introduction of tobacco in the sixteenth century changed the character of snuff. From undated times on the American continent, tobacco had been smoked through nostril tubes by the Aztecs and used solely as snuff by the Peruvians. When the plant was introduced

[1] "Toilet of Flora," 1779.

into England, its medicinal virtues to cure outward and inward ailments carried the great appeal. Although, as advertised, the smoke when inhaled cleared the head, it had a narcotic effect not found in the herbs to which people resorted as remedies for the head. On that point, the line of demarcation was made from the very beginning between tobacco as an herb for pleasure, and other herbs for medicinal use. "It is taken of some phisically in a pipe once a day at the most and at that in the morning fasting, against paines in the head, . . ."[2]

Pure tobacco was seldom used by the middle or lower classes in the early days, when it was difficult to get and expensive to buy. It was usually combined with other herbs, such as anise, fennel, sage or rosemary, of which no one would question the medicinal value. But, adulteration or not, James I of England and public opinion both in Great Britain and on the continent finally took a definite stand against the growing use of the questionable herb, and then began edicts, decrees, and laws against tobacco smoking. The old familiar herbs for head ailments again assumed the popularity enjoyed from the days of Dioscorides and Pliny. A favorite smoking mixture has always been the powdered dried leaves of coltsfoot, combined with betony, thyme, rosemary or eye-bright. "Perfume of the dried leaves (coltsfoot) layed upon quicke coles, taken into the mouth through the pipe of a funnel or tunnell, helpeth such as are troubled with the shortness of wind."[3]

During the time of the persecution of tobacco smokers, while the herb was still scarce, and while the tobacco tax was so high as to make tobacco smoking prohibitive for the lower classes, the custom of snuff taking grew up. Less tobacco was used in a few pinches of the powder than in a pipeful of the smoking mixture, and it could be taken unobtrusively. When Charles II fled to the continent after the execution of his father, he adopted many European customs which he and his courtiers brought back to England upon the Restoration in 1660. The practice of snuff taking having become almost universal in Spain, Italy and France by both men and women, it was enthusiastically welcomed in England. An expedition to Cadiz

[2] Gerard, 1597.
[3] William Coles in "Adam in Eden," 1657.

Late 18th Century Snuff Mill which Belonged to the Father of Gilbert Stuart — Still Standing Near North Kingston, Rhode Island

increased the use of snuff. Spanish ships from Havana were plundered, and the loot of fifty tons of snuff was sent to English seaports where, at a few pence a pound, it was quickly sold.

Snuff making is a slow process, the details in each century and of each manufacturer being different, though the same general rules prevail. Tobacco snuff is made of the leaves, midribs, or waste from the manufacture of smoking mixtures. The material, sprinkled with water to which salt or ammonia is added, is allowed to ferment for weeks. All snuff except that from Wales and Ireland is perfumed, this process following the first fermentation period. Tonka beans and licorice are often added, and sometimes the essential oils, rose, lavender, or bergamot, tobacco being an especially good fixative for odors. The freshly perfumed mixture is dried, ground, and allowed to ferment for a few weeks and still again, if necessary to attain the degree of scent desired. Many were the varieties of snuff, both manufactured commercially and mixed at home by individuals. In the nineteenth century, one Lord Petersham had a passion for concocting snuff mixtures, and when, after his death, the snuff was sold, it took three men three days to weigh the total amount.

The recipes in the old still-room books are a far cry from the efficient methods of the modern scientific factories. The tobacco leaves, stripped of stalks and large fibres, were dried on a mat in the sun, then ground in a mortar and sifted. A tub with a corked hole in the bottom was completely lined with a linen cloth fastened around the rim, and the snuff material was put in, covered with water and, after twenty-four hours, the liquid was strained off, and fresh water poured on. This process was repeated several times, after which the material was spread out to dry on an ozier hurdle and covered with a thick linen cloth, then again put into the tub with a "sufficient quantity" of Angelic, Orange Flower, or Rose Water. After twenty-four hours, the snuff was once more dried on the hurdle and sprinkled with scented water. But even this was not the end. The still-room mistress was now ready to perfume the mixture. The flowers that most readily communicate their flavor to snuff are Orange, Jasmine, Musk Roses and Tuberoses. A box is lined with paper and filled with alternate layers of snuff and flowers. At the end of twenty-four hours, not more, or the flowers will sour the snuff, the flowers were sifted out and discarded, and the process repeated with fresh flowers until the desired scent was attained. One suggestion for perfuming snuff with rosebuds I seized upon for my potpourri. Take rosebuds, "stripping off the green cup and pistil that rises in the middle, and fixing in its place a Clove, being careful not to separate the Leaves that are closed together. The Rose-buds thus prepared, are to be exposed to the heat of the sun a whole month, inclosed in a glass well stopped and are then fit for use."[4] Maltese, Italian and Spanish snuff were distinguished by the use of animal scents, civet, ambergris, musk, in addition to the flower scents, ofttimes with a further addition of sugar and wine, all carefully kneaded by hand.

The early method of taking snuff was through small quill-like pipes put daintily on the back of the hand and snuffed up delicately

[4] "Toilet of Flora," 1779.

Courtesy of Museum of Fine Arts, Boston

Scotch 16th-17th Century Snuff Mull — Pewter

into each nostril, one after the other. Many a discourse was written on the "exercise of the snuff box," the manner of procedure as important and significant to a man's position in society as "the exercise of the fan" was for a lady. Even into the nineteenth century, after snuff taking was no longer fashionable, costly snuff boxes were still the conventional gifts to be made to kings and potentates. Lord Petersham devoted one room to shelves and tables of expensive containers for his various mixtures and for his collection of snuff boxes, of which, according to whim, he chose a different one for every day in the year.

In Scotland, mull was the name of the snuff box, from the mill or mull, an early tobacco grinding machine. Usually, a large mull was kept on the table, and was in the form of a ram's horn with silver ornamentation, this shape derived from the fact that snuff in Scotland used to be stored in a real ram's horn.

Although the custom of snuff taking, as a social gesture, died out in the nineteenth century, snuff users, both men and women, still exist today in large numbers, though not indulging in the practice so obviously or publicly as in the past century. Today, although inhaling the powder is much done, the snuff taker can buy a rather spongy preparation, making a pellet of it to keep in the cheek, or between gum and lip. These methods are often employed by workers in factories or other places where smoking is prohibited and chewing would attract attention. Tradition has it that snuff boxes were installed in the United States Senate Chamber from its inception. The original boxes have disappeared but, as a surviving memory of the general custom of snuff taking, two modern *papier-mâché* boxes are still kept, one on the Democratic side of the house, the other on the Republican.

An unbelievable amount of snuff is still sold in this country, amounting in 1935 to the sum of $3,000,000. Most of the raw material is raised here, but, in the import figures of 1936, we find that 32,931 pounds of snuff and snuff flour were brought from the Azores, Madeira Islands, Irish Free State, Netherlands, and the United Kingdom. In 1937 there were 36,000,000 pounds of snuff sold in the United States. Mixtures of herbs, not including tobacco, are sold for burning, the fumes inhaled to relieve asthma and head ailments, and there are pastes and powders to be applied to the nostrils. But, though they were common earlier, since the world war there seem to be no herbal snuffs, as such, on the market.

To me these herbal snuffs of the old days have a fascination, in that they picture the naïve minds of our ancestors. Their knowledge of the connection between the head and the brain was elementary; their ideas on nerves and reactions, none. So it was natural that the head herbs were tried in all sorts of forms and those unscientific people were not so far wrong, since many of the same herbs are even now valued for head colds, though they make no claim to quickening the brain or restoring the memory.

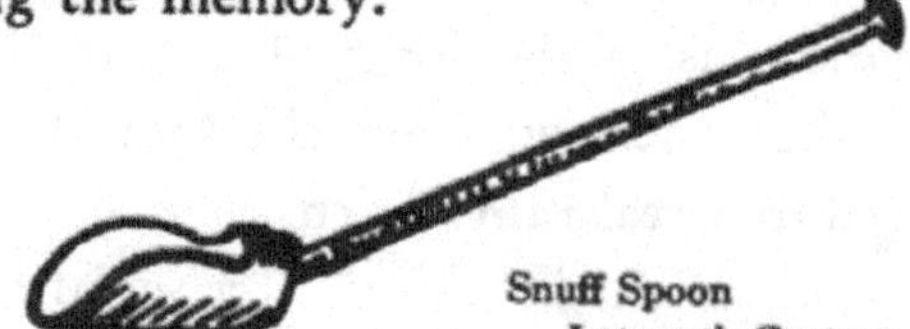

Snuff Spoon
Late 17th Century

FURTHER RESEARCHES ON EPHEDRA AND EPHEDRINE

HEBER W. YOUNGKEN

THE Chinese drug Ephedra and its active alkaloidal constituent, ephedrine, have found increasing use in the treatment of nasal colds, asthma and hay fever since their introduction into American medicine about 12 years ago, following the researches of Chen and Smith.

The Sino-Japanese trouble has reduced our supplies of Ephedra and has markedly increased the cost of the natural alkaloid. This has stimulated research in the cultivation of Ephedra in this country, with the aim of determining its possibilities as a commercial drug crop.

Attempts have been made by the writer and others to propagate *Ephedra sinica* from stem cuttings, but without success.

Prof. L. D. Hiner of the South Dakota State College, in a recent report to the American Pharmaceutical Association (August 1938), has found this species thrives well in South Dakota and that it can be successfully propagated from rootstocks and seeds of South Dakota grown plants. His plants proved both disease and insect resistant. He sowed seeds in prepared soil of open plots, the seeds being planted about 2 inches apart and from ½ to ¾ inch deep in rows 30 inches apart. These were covered with a mixture of about equal parts of sand and earth, pressed down firmly, and well watered until after germination had taken place. The plants formed a long, conical root and later sent out lateral rhizomes which developed roots and stems at their tips and, in due course, formed a dense sod. He found that by cutting around the older plants, severing the rhizomes with a sharp spade, the individual plants could be separated. These should be left in the soil and transplanted the following spring. (Exhibit of first, second and third year plants grown by Dr. Hiner.)

The aerial green stems, alone, should be harvested, the best time being in September. Hiner ascertained that the ephedrine content of his plants increased yearly, the stems from four-year plants possessing 0.818% of ephedrine. This compares favorably with the average imported article and establishes the fact that Ephedra can be grown successfully as a drug crop in South Dakota.

While lævo-ephedrine (the kind of ephedrine occurring in Ephedra) was first isolated from what was believed to be *Ephedra vulgaris* Rich by Nagai in 1887 (but which was probably *E. sinica* Stapf), it was first synthesized on an experimental scale by Späth and Göhring in 1920. Later a synthetic ephedrine appeared in commerce which proved to be the racemic variety, consisting of a mixture of dextro- and lævo-ephedrine. This has been employed at times as a substitute, but has proven to be only about half as potent as the natural alkaloid. Recently, however, a synthetic lævo-ephedrine has been made in commercial amounts and is claimed to possess the same potency as the natural article. This, however, cannot be used nor offered commercially until it receives official recognition.

ARTEMISIAS

MRS. JAY CLARK, JR.

IN the hunt for gray foliage plants of varying form and texture, a happy stumble brought me to this delightful family of many virtues and few faults, and around twenty have been gathered in, here and abroad. Of all the collection only three or four have not repaid interest and care. *Per se* they do not belong to the herb group, and yet it is one of the few plant families — I might say offhand the only one — which contributes members to the three divisional uses of the herbs: aromatic, culinary and medicinal. At first acquaintance I slipped on pronunciation, for with many gardeners it had been called ar-te-mee´-sia, while it should be sounded as it is spelled, ar-te-mis´-ia,

ARTEMISIA — PONTICA

a natural sequence from its supposed relation to the Greek goddess Artemis. The mistake is easily understood, as half the time the word is printed with a second *e* instead of the proper *i*. Legend and lore have no place here, in fact most of what we are told of the plant is factual and historical. From the carvings of the ruthless Assyrians, as interpreted by Dr. Campbell Thompson and given us in his monograph of their vegetable drugs, Artemisia's record goes back to 600 B. C. Figures of speech are always taken from familiar objects, and since Bible times wormwood has been the symbol of bitterness, for of all plants it possesses that characteristic. On the last night in Egypt, Moses had the escaping Children of Israel flavor their paschal lamb with bitter herbs, and later no matter how much they howled for the cucumbers and melons, leeks and onions and garlic they had left behind, wise Moses kept the cleansing bitterness in their diet, with his health and food laws, which even today are said to be the last word in science. From the Book of Exodus to the final Apocalypse, runs the mention of wormwood, in actuality and as the bitter symbol of futility, the purge of repentance, the curse of feckless judgment, until the frenzied star falls from Heaven upon a third of the rivers, blasting water and men from its bitterness. "And the name of the star is called Wormwood." This is the gentle plant we bring into our gardens of peace!

With rare unanimity authorities state that the artemisia of historic record was either *A. pontica,* called Roman wormwood, or *A. absinthium,* which we dub common wormwood, or both. *A. pontica* is found on either shore of the Mediterranean, is native of Southern Europe, indigenous to the Black Sea region, a difficult plant to keep in bounds. It spreads from underground runners into a lovely mass of fine feathery growth, stems not over a foot high, shrubby stalked with pale green leafage, gray underneath. It resembles *A. pedemontana,* the artemisia of Piedmont, Italy, but is twice as tall. The other probability of historic times is far better known. *A. absinthium* is an erect under-shrub, found wild in nearly every part of Europe, growing by the roadside and on heaps of old rubbish. The leaves and flowers are warm and bitter, with a strong smell; the roots are warm and aromatic, and it has always been a plant cultivated chiefly for medicinal purposes. What these are vary with the times. In the second century A. D., Dioscorides declares it to be a remedy against intoxication; Gerard in the sixteenth used it as a vermifuge, and the pharmacists recommended it for tonic properties, employing it as a stomachic. An infusion of the leaves, with the addition of a fixed alkali, is in a book of household remedies of 1854, as a powerful diuretic in cases of dropsy. Today its main uses are for the manufacture of the liqueur absinth, and various liniments. The cordial is concocted from wormwood and other flavoring ingredients as a powerful stimulant, but it has been suggested that the dangerous properties of absinth come not so much from *A. absinthium*, as from the mixture wrought into the infusion. In this country it has been put to a much tamer if more beneficent use, as an old-fashioned remedy for sprains, particularly in horses. At first it was a matter of an improvised poultice, then commercially it appeared under the name of Absorbine. Later for man's healing a more refined product appeared as Absorbine, Jr., and was long a remedy in our medicine cabinet before I had ever heard of the herb. After the war the plant was grown for this medicinal oil in large quantities in Michigan, but when the foreign imports returned in bulk, the crop was abandoned. That is the story of commercial drug growing both in this country and England: many of the plants grow wild on the Continent of Europe; the labor of harvesting them is undertaken for a small sum, and the distribution is well organized. So much for the medical artemisias.

In the aromatic division comes the plant of many names, *A. abrotanum*, Southernwood, affectionately called Old Man, Lad's Love, and regarded with a tender sentiment difficult to explain. It forms a low bush, with woody branches bearing very finely divided, pale milky green leaves that give off a strong scent when crushed. It is native to the South of Europe, Spain and Italy, and has been in England since 1548. The leaves have a fragrant, lemon-like odor, which is dependent, like all aromatics, on the volatile oil contained in the plant. The smell is supposed to be repellent to bees and moths, the reason the French call it *garde-robe* and put it among their woolens. I value it for its fragrance, a scent quite unlike that of any other herb, and for me one of the best. For those who like to run down

Artemisia — Absinthium

sobriquets, the other names of Southernwood are Boy's Love, Maiden's Ruin; the French call it Bois de St. Jean and Citronelle, the Dutch name it Averonne and the Germans Stab-wurtz. A sweetmeat was made in England from the chopped fresh leaves of Southernwood mixed to a paste with sugar, recommended for sleeplessness, and in Italy it is employed as a culinary herb, to rub on lamb before cooking in the same manner as rosemary, as well as the young shoots used for flavoring cakes.

The artemisia devoted solely to kitchen rites is tarragon, a bewildering and temperamental member of the unpredictable family. While we are intimately acquainted with but one, there are numberless others, native of Europe, North Asia and North America, each quite different from the other. The classification most familiar is Besler's, who designates the cultivated tarragon as *Artemisia dracunculus var. sativa* and the wild species as *A. dracunculus var. inodora.* X. Marcel Boulestin, writing from London in 1930, says: "The true tarragon is difficult to obtain in this country, where nurserymen almost invariably send out in place of it a nauseous plant from North America, an understudy which is probably responsible for the neglect of the herb in English cooking." *A. dracunculus sativa* is commonly though inaccurately known as French tarragon, and is the salad herb, one of the few artemisias that is not bitter, though the taste is spicy with a tinge of camphor. It is a native of Central Asia and did not get to Europe before the sixteenth century, and as it does not seem to have set seed since that time, and is intolerant of wet cold winters, the necessary propagation by division makes tarragon an uncertain quantity. The culinary herb is *not* Russian tarragon, which does seed and is not worth using. Where does the name come from? The dragon-tail serpentine coils of the roots, or the legend that it sprang up along the serpent's track in the Garden of Eden. Take your choice.

In this same culinary division should be classed *A. vulgaris,* which is the Mugwort common on waste lands. It was once a substitute for hops in the manufacture of beer, and was also made into tea in England when the imported article cost seven shillings a pound. One writer, Dr. Fernie, avers that this was the plant first prepared as a drink by a French physician named Ordinaire at the close of the

Artemisia — Dracunculus

eighteenth century, a non-alcoholic tisane quite unlike the absinth of today. One more variety of the family has a mission which puts it in the medicinal section of the herbs, *A. Cina,* native of Russia, where the wild yield provides the world with the source of santonin, an anthelmintic against the round worm. Or in domestic circles, as I can testify, it is good for what ails the kittens!

The remainder of my collection is purely ornamental, having no utilitarian motive except sheer pleasure of form and gray softness. *A. tridentata* is the sage brush of our Western plains, fragrant after a shower, whose silvery velvet sheen is lovely in the extreme. Others equally happy in like quality are *A. albula; A. gnaphalodes; A. purshiana,* the cudweed; *A. ludoviciana,* white sage, and the European *A. maritima,* very like our beach wormwood, but taller. These are all of 2-3 feet growth. Lower plants of the same gray silver are *A. Stelleriana,* also called Old Woman and Dusty Miller, which grows on the stretches of the North Atlantic in pure sand, and is yours for the taking; *A. frigida,* fringed wormwood, whose silvery gray foliage is very finely cut, the woody stems holding it above the ground; *A. argentea,* whitest of all and delightful in sheen, the leaves fine threads of shining silk, and *A. spicata,* good for cutting and eighteen inches tall. The low ground huggers would have been suitable for a rock garden, had I possessed one, but they were quite happy in the sandy pockets of Artemisia Allée. Such were *A. Baumgartenii; A. glacialis; A. stellata,* finely cut foliage, fragrant, *not* to be confounded with *A. stelleriana; A. nana; A. granatensis; A. pedatifidia.* Then there are the annuals, *A. sacrorum* or Russian Wormwood, much used for moth repellent, and *A. sacrorum viridis,* Summer Fir, a bold, handsome thing of weed-like growth, and *A. annua* of no virtues I can discover! *A. lactiflora* should not go unmentioned, although it is a lottery — very lovely when the real Chinese plant is obtained; often, however, *A. vulgaris* is sent in its stead. All in all it is a fascinating family, whether it is approached through the door marked Herbs or merely the one that indicates Pleasure of Possession.

A NOTE ON THE FRONTISPIECE

G. M. BROWN

AS far back as the days of the Romans, and in all early herbals, we find mention of catnip as an important medicinal herb used in the treatment of a long and varied list of ailments. In 1620, Captain John Mason included it among the eleven herbs he considered essential for a Newfoundland fisherman's garden. It was planted in Salem in the early days of the colony, and was widely used medicinally, as well as for a soothing tisane. Catnip is still listed as an official drug, consisting of the dried leaves and flowering tops of Nepeta Cataria. Although a certain amount is imported, most of the supply sold in this country is home grown, and is used in the form of a 5% infusion, chiefly as a carminative. Infusion of catnip is an old remedy, and there is also the official National Formulary Elixir of catnip and fennel, a frequently prescribed medicine as an agent to induce perspiration in fever. While nowadays the medical demand is not as general, the thriving trade in this dried herb is a proof of the continued tradition of the exhilarating effect it has on the feline race. There are still catnip farms and catnip venders. These peddlers, although fast disappearing, can be found here and there in some of our New England seacoast towns. Our frontispiece shows a familiar Boston figure, Richard Beaudreau, who for twenty-nine years has stood in front of the old Granary Burial Ground with his basket.

Literature References

For " Colchicine " by Mabel L. Ruttle

Blakeslee, A. F.

1937. *De doublement du nombre de chromosomes chez les plantes par tratement chimique.* C. R. Acad. Sc. 205: 476-479.

Blakeslee, A. F., and Avery, A. G.

1937. *Methods of inducing doubling of chromosomes in plants.* Jour. Heredity 28: 393-411.

Nebel, B. R.

1937. *Cytological observations with colchicine.* Collecting Net 12: 130-131 and Biol. Bull. 73: 351.

Nebel, B. R., and Ruttle, M. L.

1938*A*. *The cytological and genetical significance of colchicine.* Jour. Heredity 29: 1-9.

1938*B*. *Use of chemicals in plant breeding opens up new possibilities.* Farm Res. 4(3) July.

1938*C*. *Colchicine and its place in fruit breeding.* N. Y. Agr. Exp. Sta. Circular No. 183 (In Press).

Ruttle, M. L.

1938. *Some common mints and their hybrids.* The Herbarist 4: 17-29.

Ruttle, M. L., and Nebel, B. R.

1938. *Cytogenetic results with colchicine.* Biol. Centralblatt (In Press).

This article has been approved by the Director of the New York State Agricultural Experiment Station as journal paper 309, January 16, 1939.

Wenham Exchange

PRESENTS

MARY CHESS

of New York and London

PERFUMES - - SACHETS
TOILET WATERS
BATH LUXURIES

WENHAM MASSACHUSETTS
Telephone: Hamilton 235-W

INTERESTING
SOUTHERN DISHES

Home cooked and vacuum sealed in modern sanitary containers.

A few of the items are:

WHOLE BROILERS with Pecan stuffing.

CHARLESTON CHICKEN PILAU, a rice dish with almonds and raisins and curry.

HICKORY SMOKED NUGGETS, tenderloin cooked like baked ham, with spiced fruit syrup and raisins.

STEAMED CAROLINA FRUIT LOAF

PECAN TEA-BREAD

Sold only by mail, direct from the kitchens. Send for price list and kindly mention **The Herbarist.**

RALPH ERSKINE
Pacolet Plantation
Tryon, North Carolina

G. S. Cheney Co.

15 UNION STREET
BOSTON, MASS.

For almost a century G. S. Cheney Co. has dispensed herbs, drugs, pharmaceuticals and chemicals.

They invite all herb lovers to visit their permanent exhibit

HOVÉ

PARFUMEUR

Perfumes of the Highest Quality

At Moderate Prices
Specializing in Flower Scents

723 Toulouse Street
"Casa Hové," near Royal Street
(Formerly at 529 Royal Street)
New Orleans, La.

THE HERB FARM SHOP OF LONDON

Comes to America!

Photograph shows the Herb Farm Shop at Bonwit Teller, New York City

• Herbal preparations of rare goodness, a traditional part of English life and loveliness, are now available to the American gentlewoman at Bonwit Teller in New York City and in other selected agencies throughout the States.

• Here you will find the same flavoring herbs, jellies, herb vinegars and teas . . . herbal soaps, creams, lotions, bath essences and potpourri . . . beloved by members of the Royal Family and other esteemed patrons of our London shop at 16, North Audley St.

THE HERB FARM SHOP OF LONDON, Ltd.

IN ENGLAND:
16 NORTH AUDLEY STREET, LONDON, W 1

IN AMERICA:
347 FIFTH AVENUE, NEW YORK

Old English Herbs from the Olitory at Stonehurst, Shanty Bay, Ontario, Canada.

Balm	Parsley
Basil	Sage
Dill (Seeds)	Summer Savory
Fennel	Tarragon (Leaves)
Marjoram	" (Ground)
Mint	Thyme

The Herbs furnished from the Olitory at Stonehurst, may be used dry at the table to flavour soups, sprinkle potatoes, etc., the same as you use pepper and salt.

These herbs are grown in suitable soil, in sunshine or shade as is most fitting, harvested at the right time, dried in the proper way, carefully winnowed and finely powdered. They are absolutely pure and much stronger flavoured than herbs usually obtainable.

1 Jar....................	$ .50
Box of 6 jars............	2.50

These herbs make an attractive and useful Christmas present. A "Herbal" of historical value and containing many receipts from the old herbals, beginning with "Dioscorides" of the first century A.D. up to the XVIII century, is sent with each box.

Toronto Office

THE GREENMAN

184 Bay Street Elgin 2953

Basil

Mint

The Greenman

Sage

Thyme

Made in the USA
Monee, IL
13 September 2026

62981003R00164